Studies in Natural Language and Linguistic Theory

VOLUME 45

The titles published in this series are listed at the end of this volume.

BEYOND PRINCIPLES AND PARAMETERS

Essays in Memory of Osvaldo Jaeggli

edited by

KYLE JOHNSON

University of Massachusetts, Amherst, U.S.A.

and

IAN ROBERTS

University of Stuttgart, Germany

KLUWER ACADEMIC PUBLISHERS

DORDRECHT / BOSTON / LONDON

Library of Congress Cataloging-in-Publication Data

Beyond principles and parameters : essays in memory of Osvaldo Jaeggli / edited by Kyle Johnson and Ian Roberts.
p. cm. -- (Studies in natural language and linguistic theory ; v.45)
Includes bibliographical references and index.
ISBN 0-7923-5498-2 (alk. paper)
1. Grammar, Comparative and general--Syntax. 2. Generative grammar. I. Jaeggli, Osvaldo. II. Johnson, Kyle, 1958- III. Roberts, Ian G. IV. Series.
P291.B49 1999
415--dc21 98-47895

ISBN 0-7923-5498-2

Published by Kluwer Academic Publishers,
P.O. Box 17, 3300 AA Dordrecht, The Netherlands.

Sold and distributed in North, Central and South America
by Kluwer Academic Publishers,
101 Philip Drive, Norwell, MA 02061, U.S.A.

In all other countries, sold and distributed
by Kluwer Academic Publishers,
P.O. Box 322, 3300 AH Dordrecht, The Netherlands.

Printed on acid-free paper

Printed in the Netherlands.

TABLE OF CONTENTS

INTRODUCTION

Kyle Johnson
University of Massachusetts at Amherst

Ian Roberts
University of Stuttgart

An important chapter in the history of syntactic theory opened as the 70's reached their close. The revolution that Chomsky had brought to linguistics had to this point engendered theories which remained within the grip of the philologists' construction-based vision. Their image of language as a catalogue of independent constructions served as the backdrop against which much of transformational grammar's detailed exploration evolved. In a sense, the highly successful pursuit of phonology and morphology in the 19th century as compared to the absence of similar results in syntax (beyond observations such as Wackernagel's Law, etc.) attests to this: just noting that, for example, French relative clauses allow subject-postposing but not preposition-stranding while English relatives do not allow the former but do allow the latter does not take us far beyond a simple record of the facts. Prior to this point, syntactic theory had not progressed beyond the 19th century situation.

But as the 80's approached, this image began to give way to a different one: grammar as a puzzle of interlocking "modules," each made up of syntactic principles which cross-cut the philologist's constructions. More and more, "constructions" decomposed into the epiphenomenal interplay of encapsulated mini-theories: $\overline{X}$ Theory, Binding Theory, Bounding Theory, Case Theory, Theta Theory, and so on. Syntactic analyses became reöriented toward the twin goals of identifying the content of these modules and deconstructing into them the descriptive results of early transformational grammar.

Emboldened by the example of John Ross's dissertation, Chomsky and his students began to peel away those elements common to the sets of rules which, in transformational grammar, defined constructions. This process led eventually to the image of encapsulated modules described above.

Consider an example from Bounding Theory.

After Ross's dissertation, it became increasingly clear that many "unrelated" constructions in English were subject to the same or similar constraints. The anaphor-antecedent relation, for example, is apparently subject to a locality condition rather like that which holds of "Argument Movement." This locality condition blocks the

K. Johnson and I. Roberts (eds.), Beyond Principles and Parameters, 1-11.

movement and the anaphor binding relations from holding across Subjects, as the examples in (1) illustrate.

(1) a. *Phil$_1$ believes [$_S$ **Loretta** to like himself$_1$].
b. *Phil$_1$ was wanted [$_S$ for **there** to be t_1 in the garden].

The phrases *there* and *Loretta* are Subjects between the terms which are associated by the rules involved (here signaled by the use of identical indices); and for this reason, the sentences fail. Chomsky attempted to extend this locality condition so that it held of every rule which establishes a (syntactic) discontinuous dependency. This led to the formulation of the constraint in (2).

(2) No rule can involve X, Y in the structure:
.... X .... [$_S$ ... Z ... WYV ...] ....
where Z is the specified subject of WYV.
(Chomsky (1977), p. 90)

And this constraint eventually evolved into a central component of Binding Theory, the module which governs those relations represented with "referential indices."

It's this method of factoring out elements common to the grammars of constructions which has led to the modern modular arrangement of syntactic theory. Syntax is conceived of as a series of independent sets of principles whose interaction yields the complex syntactic behavior of individual languages, including their constructions. Language differences can then be identified with open parameters within those principles – parameters which, once fixed, give final form to the language and their constructions. This is the Principles and Parameters framework, which now serves as the backdrop for most of the theorizing done in modern syntax.

While the Principles and Parameters framework got started as a deconstruction of constructions, it got its big push from sustained, detailed comparative work. Through the efforts of Richard Kayne, Paul Postal, David Perlmutter, Luigi Rizzi and many others, it became ever clearer that a construction-based view of grammar does not give an accurate portrayal of where language variance and linguistic universals are to be found.

Moreover, comparative work provided a powerful tool for uncovering the otherwise hidden composition of these modules. Language comparison allowed one to isolate the dimensions through which constructions varied, and this projected a surprisingly precise image of the principles that they are constructed from. Comparative work thereby delivered the tools necessary to build new theories at the same time that it revealed the fatal flaw in the old ones. The lectures Chomsky delivered in Pisa in 1979 were heavily influenced by this new perspective, and in turn helped focus the attention of much of the syntactic community on the project that remains the mainstay of modern syntax.

Kayne's (1975) important study of French, for example, demonstrated, among other things, that the locality constraint in (2) is operative in a construction which English does not house. This construction is brought about by a rule of Clitic Placement, a rule which when combined with a general system of word-order, determines whether or not an argument will appear displaced into the verbal nexus. When arguments are "weak pronouns," they will be so displaced; otherwise they won't. Kayne (1975, p. 287f.) discovered that this rule is subject to (2), which is the cause of (3)'s ungrammaticality.

(3) a. *Elle$_2$ te$_1$ veut [$_S$ PRO$_2$ voir t_1].
She you wants see.

b. *Elle$_2$ te$_1$ décidera de [$_S$ PRO$_2$ voir t_1].
She you will-decide to see

Partly on this basis, Kayne argues for a movement dependency between clitics and their grammatical-function position, bringing Clitic Placement under the general schema in (2).

From comparative work of this sort, we learned that not only is a construction-based view of grammar not able to capture generalizations of the sort demonstrated above, but, perhaps more surprisingly, neither is a language-based grammar. The limit on long-distance dependencies that (2) aims at doesn't even respect language's borders. It respects, instead, only the seams at which the parts of Universal Grammar are joined. In this example, then, we see how comparative syntax frees us from the blinding imposition of constructions, and even the hoary sociological constructs of language boundaries, and lays bare grammar's parts.

When the comparative method of carving up Universal Grammar is pursued in earnest, "Language" seems to break apart into domains, each with its own set of shaping principles and constraints. Some of these domains conform to the outlines of the modules discussed above. Anaphora systems, for instance, make use of principles largely encapsulated in the Binding Theory module. But other domains seem to lie wholly within these modules. Such is the case, for instance, with Bounding Theory. Here we find that there are classes of movement – Argument Movement, Head Movement, and so on – each distinguished, roughly, by the qualities of their landing sites. These movements appear to be subject to somewhat different sets of bounding constraints, and to interact in somewhat different ways with other processes. Still other domains seem to cross-cut the modules. A salient example is the principles which govern how θ-roles are related to their bearers throughout the derivation (the Projection Principle), or, more generally, the principle(s) that determine how operations are timed throughout the derivation (the Principle of the Cycle).

Moreover, language comparison also reveals that these domains do not preserve their integrity cross-linguistically. Clitic Placement offers a good example. By virtue of the properties of the landing site, Clitic Placement shares traits with those processes that partake of Head Movement. Clitics appear to be X°s adjoined to other X°s, just as in the canonical exemplar of Head Movement: Verb Movement. Yet, Clitic Placement is subject to a locality condition that better mimics Argument Movement; it does not seem to be subject to the tighter locality condition that Verb Movement, for instance, obeys. Thus, the properties that line up so neatly in English (and, in this respect, the other Germanic languages too), pointing to a tight connection between the "kind" of movement involved and the locality constraints invoked, begin to fall apart in the Clitic construction.

In short, comparative syntax is beginning to unravel the integrity of the modules that the Principles and Parameters framework delivered.

A similar historical trajectory can be witnessed in theories of word-order and the representation of grammatical functions. As the construction-based grammars of Passivization, Raising, Dative Shift and similar phenomena began to give way to a decomposition into the modules of Theta Theory and Movement, it became clear that Case assignment was a key factor in determining where "subjects," "objects" and "indirect objects" are placed. On the basis of English data, it is nearly possible to determine completely where these arguments will be found in surface representations on the basis of their Case morphology alone. But with a modest amount of language comparison, it became clear that this is rare. The position "objects" take appears to be connected not only to their Case morphology in many languages, but also to various of their semantic properties, as well as to their syntactic form. Similarly, the position "subjects" assume is connected not just to their Case morphology, but to properties of the predicates they are subjects for (among other things). Case is just one of a number of factors involved in giving arguments their surface position.

Now, this is an awkward situation for the Principles and Parameters model. It indicates that the modules identified at the outset do not match up with the divisions which emerge from cross-linguistic comparison; they provide too coarse an image of Universal Grammar. Moreover, it has been suggested that language variation is wholly bound up in the mix of properties, or features, that are increasingly implicated in syntactic operations. Thus, the vision of UG as clusters of principles with open parameters, is beginning to blur.

In the last five years or so, it's this problem that has driven a good deal of the research in theoretical syntax. The search is on for a decomposition of the modules into smaller systems that more accurately draw the divisions among syntactic phenomena. A recurrent ingredient in much of this work is the introduction of what might be called, "proprietary context-sensitivity." In Rizzi (1990), for instance, the slightly different bounding constraints found in the various sorts of movement operations are unified into one "relativized" locality constraint. This constraint factors out the common syntactic environment that all of the local constraints make reference to, and targets a particular blocking term which triggers the constraint. This item is defined contextually, in terms of properties it shares with the moved item. The constraint's action therefore changes with the kind of movement involved. Chomsky's (1986) formulation of the locality constraint found in the Binding Theory has a similar property. It frames this locality constraint in terms of properties held by the item involved. Personal pronouns will be subject to a different locality constraint than will anaphors, therefore, not because they are subject to different constraints, but because they instantiate different binding theoretic properties to which the constraint they obey is sensitive. The trend, then, has been to recraft the sets of principles into general constraint schemas, whose satisfaction is very narrowly contingent on properties of the items involved in the constrained process.

In the case of word-order, this trend is married to two other developments: a reworking of the syntax of argument projection, triggered by Larson's (1988) study of the double object construction, and a calculus of feature "strength," which is intended to determine when the satisfaction of constraints yields visible results or not. Chomsky and his colleagues have, in the last few years, tried to organize these trends and developments into a framework: the so-called Minimalist Program, which makes heavy use of proprietary context-sensitivity. The future of these trends, and this program, remain wide-open, many of the fundamental problems uncovered by comparative work yet to be solved.

One researcher whose contributions played a central role in the consolidation of the Principles and Parameters approach was Osvaldo Jaeggli. His 1980 dissertation,[1] the first from MIT to take full advantage of the shift in orientation, examined the syntax of clitics, and the conditions on empty categories which govern the distribution of PRO, trace and what was to become known as *pro* (the null pronominal which instantiates null subjects in languages like Italian). The issues he addressed and the ideas he pursued foreshadowed much of what preoccupied syntacticians throughout the 1980s. His formulation of the Empty Category Principle, for example, is, at its core, a preliminary version of the sophisticated ECPs which Lasnik and Saito have arrived at, and which, in turn, have fostered Rizzi's work.[2] His analysis of clitic placement provided an alternative to the movement account of Kayne (1975) mentioned above, and similarly guided work on cliticization in the decade that followed. His subsequent work on the complementiser-trace phenomenon, *to*-contraction, null subjects, passives and the inflectional behavior of aspectual verbs has had, in each instance, a shaping role on the research carried out on these topics.

The authors assembled here are among those influenced by Osvaldo's work, and have themselves played a role in the transition from construction-based to principles-based grammars. The work they report in these papers takes important steps beyond the Principles and Parameters framework, and, in many cases, touches directly on the legacy that Osvaldo left us. The subjects treated here range over those identified above as instrumental in making the transition from the Principles and Parameters vision.

For example, the reworking of the syntax of argument projection, which has played a role in the modernization of word-order theories, is the subject of Anne Rochette's contribution. She provides an analysis of aspectual verbs like *begin, finish,* and *start* that explains how they are able to invoke the apparently different argument structures illustrated in (4).

(4) a. The noise began to annoy Joe.
b. John began the letter.

In (4a), *begin* behaves like a raising verb, assigning no external θ-role but combining with a nonfinite complement whose subject raises to serve as its surface subject. But in (4b) it apparently functions as a transitive verb, taking both an external and internal argument. As with many such alternation's, the long-standing puzzle (since at least Perlmutter (1970)) has been to understand why verbs of this class systematically permit both frames.

Rochette argues that the alternation in (4) is only apparently one of argument structure. Instead, she argues that in both (4a) and (4b), *begin* functions as a raising predicate, the difference in them residing solely in the categorial nature of their complement. Using the typology of events that Pustejovsky (1988) sketches, she argues that in both (4a) and (4b), *begin* selects a process whose external argument in both cases is the surface subject of *begin*. The alternation in (4) then is very much like that in (5).

(5) a. Jane asked what time it is.
b. Jane asked the time.

Just as the complement to *ask* in (5b) is a concealed question, whose canonical realization would be the clause in (5a), so is the complement to *begin* in (4b) the concealed process that is more fully revealed in (4a).

A couple of the articles build on Osvaldo's contribution to our understanding of the clitic construction, and are aimed at recent attempts to resolve the paradoxical mix of properties, outlined above, that this construction seems to present.

Dominique Sportiche's paper provides an account of Complex Inversion and Subject Clitic Inversion in French which relates these constructions to a conception of clitics found in Sportiche (1992), and embeds parts of the approach to clitics that Jaeggli (1982) proposed into recent advances in the syntax of Case and agreement phenomena. In Sportiche (1992), an analysis of Romance object clitics is provided under which clitics are functional categories that head phrases whose Specifiers host silent pronouns (*pro*). This explains how clitics are able to at once have properties of Head Movement and of Argument Movement. In this paper, Sportiche extends this account to subject clitics, which yields a picture of subject cliticization that diminishes the otherwise apparent differences between French subject clitics and the subject clitics of Italian dialects. He then argues for a fundamental difference between pre-verbal subject clitics and post-verbal subject clitics that takes advantage of Chomsky's (1992) proposal that certain verbal affixes should be seen as attached to the verbal stem before the syntax. He shows that unlike preverbal clitics, postverbal clitics should be analyzed much as agreement morphemes are in Chomsky's framework; they are morphological parts of the verb, rather than words adjoined to the verb. This not only allows for an understanding of a host of otherwise puzzling differences between pre and post-verbal clitics, but also leads to a very simple account of clitic Inversion constructions that form the focus of the paper. The key innovation in Sportiche's account of Complex Inversion and subject clitic Inversion in French is in departing from the standard conclusion that these constructions involve overt I to C movement. Instead, he suggests that this movement is covert, accounting for the restriction of these constructions to root contexts, but also allowing enough room between the fronted *wh*-phrase that is found in Complex Inversion and the verb for the subject to be sandwiched – a feature of the construction which has led previous analyses into considerable complexity.

Joseph Aoun's paper makes an interesting companion to the account of clitics sketched in Sportiche's paper. Working from Lebanese Arabic, Aoun tackles more narrowly the peculiar phenomenon of clitic doubling. He makes the interesting discovery that in Lebanese Arabic, the doubled argument is not in the same structural position that non-doubled arguments are. Using facts about the scopes that quantificational arguments may have when they are part of a clitic doubling construction and when they are not, and correlating these facts with the ability of these arguments to bind pronouns, he argues that the term which a clitic doubles is actually in a subject relationship to the predicate containing the clitic. Because a clitic-doubled argument always follows the clitic, this has interesting consequences when considered in the light of Kayne's Antisymmetry proposal. One of these, that the predicate containing the clitic must have fronted past the doubled argument, Aoun argues to be correct. More generally, as Aoun points out, when this analysis of clitic doubling in Lebanese Arabic is considered in the light of its differences with clitic doubling in, say, the Romance languages, we are reminded that a satisfactory typology of these phenomena are still in the future.

Many of the papers here touch on the task of reworking word-order typology, with a particular emphasis on identifying the conditions beyond Case morphology that control where arguments will be placed in surface representations.

Maria Luisa Zubizarreta's article, for example, examines the different position that clausal subjects have in Spanish, when compared with the positions that subjects may occupy in Italian and French. Her paper touches on several of the same issues discussed in Sportiche's treatment of French Complex Inversion. Together, the two papers sketch an interesting typology of Romance post-verbal subjects.

Zubizarreta shows that postverbal subjects in Spanish are not as low in the phrase-marker as are postverbal subjects in Italian and French. She provides an account of this difference that correlates it with the freedom of word-order that Spanish enjoys relative to Italian and French. Building on Roberts (1993), she suggests that the functional projections that fix Case assignment and agreement divorce nominative Case assignment from subject agreement in Spanish. In particular, she shows that Spanish differs from Italian and French in having a functional projection responsible for assigning nominative Case that lies above the functional projection that expresses subject agreement. It's the presence of this functional projection which allows nominative Case-marked subjects to have the Romance-exceptional positions that they do in Spanish. Further, Zubizarreta argues that the presence of this extra functional projection plays a role in generating apparent Verb Second word-orders in Spanish. The Specifier of this functional projection can be occupied by adjuncts, or act as the landing site for fronted arguments, yielding in certain instances XP^V^Subject arrangements. Her proposal, then, connects a cluster of properties of Spanish word-order to the mechanism underlying nominative Case assignment that is mirrored in recent work on Germanic cross-linguistic variation.

The job of finding the principles which govern the surface position of arguments in clauses, towards which these papers contribute, has been made more exciting by the conjecture popularized in Steven Abney's dissertation that the organization of nominal expressions is parallel to that of clauses. This conjecture generally credits nominals with functional projections into which the NP is buried, in just the same way that VPs are thought to be buried within inflectional phrases in clauses. This conjecture, which goes by the name of the DP Hypothesis, is used in several of the papers as a means for revealing the relationship between functional architecture and word-order.

The contribution by Saito and Murasugi explores the parallelism between clauses and nominals that the DP Hypothesis expresses, and credits a set of apparent differences between these two to general conditions on ellipsis and movement. To begin with, they argue that like the well-known ambiguity between control and raising that verbs like *promise* display (see (6)) – an alternation related to that discussed by Rochette in (4) – I° should be considered capable of setting up a control or raising relation to the underlying subject position within VP, as diagramed in (7).

(6) a. John promised to work.
 b. This analysis promises to work.

(7) a. $DP_1\ I^\circ\ [_{VP}\ t_1\ \overline{V}]$
 b. $DP_1\ I^\circ\ [_{VP}\ PRO_1\ \overline{V}]$

Embracing this conclusion captures a variety of conditions under which VPs may elide and front. In particular, the identity conditions on VP ellipsis and Fiengo's Proper Binding Constraint will apply correctly to fronting and ellipses constructions, segregating the bad from the good, under the hypothesis that the subject position within VP may be occupied by either a trace or PRO.

Interestingly, the pattern of VP ellipsis and fronting that can be derived from the PRO/trace alternation is significantly different from that which NP ellipsis and fronting display. This lures one to the belief that the underlying subject position within DPs is not similarly occupied by trace or PRO; that is, that the options for DPs in (8), parallel to (7), are not found.

(8) a. $DP_1\ D^\circ\ [_{NP}\ t_1\ \overline{N}]$
 b. $DP_1\ D^\circ\ [_{NP}\ PRO_1\ \overline{N}]$

Saito and Murasugi lead us away from this temptation, showing first that the PRO/trace alternation does play a role in DPs, and that the differences between DPs and clauses stem from how a head government requirement is satisfied.

Montalbetti's contribution also exploits the often difficult task of distinguishing PRO from traces. He reveals that the class of verbs whose indirect objects appear able to move into subject position under the Passive are actually arguments of infinitival complements. Thus, whereas these cases appear to have the structure sketched in (9), they are actually composed as in (10).

(9) IO_1 $V_{Passive}$ t_1 $CP_{Infinitive}$
(10) $SUBJ_1$ $V_{Passive}$ $[_{IP}$ t_1 VP]

He makes further interesting connections between this construction and the naked infinitives that function as complements to perception verbs in English.

Hilda Koopman offers an account of the internal syntax of pronominal DPs that capitalizes on exploding the DP Hypothesis so that it includes a series of functional projections along the lines that Pollock (1989) proposes for clauses. Like Ritter (1988) Bernstein (1991; 1993) and Picallo (1991), she sees DPs as being made up of a Number Phrase and Person Phrase as well as DP and NP. As in Ritter (1995), she argues that pronouns can be fit into this articulated structure in one of several ways – either as D° (as in Postal's (1969) famous proposal) or as part of the Number Phrase that D° embeds. This proposal is fueled by the elaborate pronoun systems that many languages host – the object of her study is Welsh, whose pronoun series she shows can be seen as various combinations of the options expressed above.

Koopman then goes on to show that the placement of pronouns within clauses, differing as it often does from the placement of full nominals, can be at least partly seen as a consequence of their internal composition. Again working from Welsh, she shows that her proposals about the internal design of the various pronoun series can be used to explain why certain of these pronominals trigger agreement while others don't. She extends this idea to give an account for the well-known differences between pronoun placement and full DP placement in Scandinavian, as well as for lesser known, and recalcitrant, problems in characterizing the placement of English pronouns. Her paper opens a new dimension on the puzzling factors which drive Argument Movement.

Hagit Borer's paper is also on the structure of DPs, but concentrates on the internal arrangement of constituents. Her work place is in the differences between the free and construct state nominals as they cross-cut with those of process and non-process nominals in Hebrew. One intriguing property of both construct and free state DPs in Hebrew is their rigid head initialness. A popular account of this, at least since Ritter (1987) (and see Ritter (1988)), has involved Noun Movement in much the same way that rigidly Verb Initial languages are often defined in terms of Verb Movement (as in Emonds (1980), for example). In both cases, we might see the NPs and VPs as having an underlying Specifier^Head^Complement word-order that is obscured by the ubiquitous movement of the head leftwards. Indeed, Kayne's recent (1994) suggests that this is the only way head initial word-orders can be manufactured.

Borer supports this characterization of the Noun initialness of DPs in Hebrew for process nominals. These, she argues, are constituted of VPs embedded within a nominalizing N° morpheme, and involve a leftwards movement of the V as a consequence. Hence, the Noun initialness of these DPs arises not, strictly speaking, through movement of the N°, but rather through movement of a V° within the DP -- a movement that creates from this V° the noun. Non-process nominals, by contrast, she argues are head initial underlyingly; they are, moreover, underlyingly N^Complement^Specifier, a violent counterexample to the typology of Head Initialness that Emonds and Kayne envision. Her argument is based on a class of word-order differences between process and non-process nominals which can be made sense of if only one of them involves Head Movement. She identifies a separate process of Noun Movement present in the construct state, one which brings the head noun and its complement into a "quasi-Incorporated" state. And she offers a variety of other proposals about the construct state which derive, and correlate, some of its peculiar properties.

With Osvaldo's death we lost one of the architects of modern theoretical syntax. This book is a small marker of our indebtedness to his work, but also a tribute to his personal influence on us as well. He was a powerful teacher, a terrific colleague and always a good friend.

NOTES

1 Published as Jaeggli (1982).

2 See Lasnik and Saito (1992).

REFERENCES

Abney, S.: 1987, *The English Noun Phrase in its Sentential Aspect,* Doctoral Dissertation, Massachusetts Institute of Technology.

Bernstein, J.: 1991, 'DPs in French and Walloon: Evidence for Parametric Variation in Nominal Head Movement', *Probus* **3**, 101-126.

Bernstein, J. B.: 1993, *Topics in the Syntax of Nominal Structure Across Romance,* Doctoral Dissertation, City University of New York.

Chomsky, N.: 1973, "Conditions on Transformations." In S. Anderson & P. Kiparsky, eds., *A Festschrift for Morris Halle,* New York: Holt, Rinehart & Winston, 232-286. [reprinted in Chomsky, N.: 1977, *Essays on Form and Interpretation*, Amsterdam: North Holland, pp. 81-162.]

Chomsky, N.: 1981, *Lectures on Government and Binding.* Dordrecht: Foris.

Chomsky, N.: 1986, *Knowledge of Language*, Praeger Publications.

Chomsky, N.: 1995, *The Minimalist Program.* Cambridge, Mass.: MIT Press.

Emonds, J.: 1980, 'Word Order in Generative Grammar', *Journal of Linguistic Research* **1**, 33-54.

Jaeggli, O.: 1982, *Topics in Romance Syntax*. Foris Publications, Dordrecht Holland.

Jonas, D. and J. D. Bobaljik: 1993, 'Specs for Subjects: The Role of TP in Icelandic', *Papers on Case and Agreement I.* In J. D. Bobaljik and C. Phillips, eds., Massachusetts Institute of Technology, pp. 59-98.

Kayne, R.: 1975, *French Syntax: the Transformational Cycle*. MIT Press, Cambridge, Massachusetts.

Kayne, R. S.: 1994, *The Antisymmetry of Syntax*. MIT Press, Cambridge, Massachusetts.

Larson, R.: 1988, 'On the Double Object Construction', *Linguistic Inquiry* **19**, 335-392.

Lasnik, H. and M. Saito: 1992, *Move α*. MIT Press, Cambridge, Massachusetts.

Perlmutter, D.: 1971, *Deep and Surface Structure Constraints in Syntax*. New York: Holt, Rinehart & Winston.

Picallo, C. M.: 1991, 'Nominals and Nominalizations in Catalan', *Probus* **3**, 279-316.

Pollock, J.-Y.: 1989, 'Verb Movement, UG and the Structure of IP', *Linguistic Inquiry* **20**, 365-424.

Postal, P. M.: 1969, 'On so-called 'pronouns' in English', In D. Reibel and S. Schane, eds., *Modern Studies in English*, Prentice-Hall, Englewood Cliffs, New Jersey.

Rizzi, L.: 1982, *Issues in Italian Syntax*. Dordrecht: Foris.

Rizzi, L.: 1990, *Relativized Minimality*. MIT Press, Cambridge, Massachusetts.

Ritter, E.: 1987, 'NSO Orders in Modern Hebrew', In J. McDonough and B. Plunkett, eds., *North Eastern Linguistics Society*, Graduate Students Linguistics Association, pp. 521-537.

Ritter, E.: 1988, 'A Head-Movement Approach to Construct State Noun Phrases', *Linguistics* **26**, 909-929.

Ritter, E.: 1995, 'On the Syntactic Category of Pronouns and Agreement', *Natural Language and Linguistic Theory* **13**, 405-443.

Roberts, I.: 1993, *Verbs and Diachronic Syntax*. Kluwer Publications, Dordrecht, The Netherlands.

Ross, J. R.:1967, *Constraints on Variables in Syntax*. Doctoral Dissertation, MIT.

CLITIC-DOUBLED ARGUMENTS*

Joseph Aoun
University of Southern California

1. INTRODUCTION

The purpose of this paper is to investigate the phenomena which have been regrouped under the titles of 'clitic doubled constructions', 'clitic doubling', or 'argument doubling'. These phenomena have been the subject of extensive discussions in contemporary linguistic studies. I believe that their interest ultimately resides in the fact that they open a window on the formation of arguments; they allow us to better understand how argument structures are formed and how they are syntactically projected.

In sentences involving doubling, the complementary distribution between clitics and lexical complements breaks down: the clitic and the lexical element co-occur:

(2) .... head + clitic .... lexical NP

Some of the main questions in this context focus on the relation between the clitic or the complex head+clitic and the lexical NP and on the status of this doubled NP. What is the nature of this relation? Is it a head-complement relation? Is the lexical NP generated in the same argument position regular complements are generated in? It comes as no surprise that the possible answers to these questions have been put forward in the relevant literature (see Aoun (1979, 1981), Borer (1983, 1986), Dobrovie-Sorin (1990), Everett (1984), Franco (1991), Hurtado (1984), Jaeggli (1982, 1986), Iatridou (1991), Schneider-Zioga (1990, 1993), Sportiche (1992), Suñer (1988, 1992)).

In this work, I propose a novel analysis of doubling based on the working of this phenomenon in Lebanese Arabic (henceforth, LA). I suggest that doubling in this language is to be viewed as a means for forming new functional structures. Specifically, I argue that in LA, doubled elements are not generated in complement positions. Rather, the doubled element is to be analyzed as a subject in a complex predication relation holding between the doubled element and the minimal complete functional complex (CFC) containing the clitic. In this predication relation, the clitic-*pro* complex, i.e. the clitic and the non-overt pronominal coindexed with it, plays the role of predicate-variable. It is the open position which makes the CFC a one-place predicate:

(3) .... [[$_{CFC}$ [.... clitic-*pro* .....]] doubled element]

K. Johnson and I. Roberts (eds.), Beyond Principles and Parameters, 13-42.

Unlike standard subjects in LA, the doubled element always appears to the right of the predicate it is associated with, and c-commands the arguments contained within this predicate, thus, apparently violating the Linear Correspondence Axiom of Kayne (1994). I will provide evidence showing that doubled elements, like other subjects in LA, are underlyingly generated to the left of the predicate they are construed with. Furthermore, I will show that the relation between the doubled element and the pronominal clitic is subject to the standard disjointness effect applying to pronominals. This disjointness effect highlights the non-complement status of doubled elements and distinguishes them from non-doubled elements. In this account, doubling is a means LA uses to create non-standard argument relations, to superimpose new functional structures on conventional ones.

2. WORD-ORDER IN LEBANESE ARABIC

I will start by providing some information regarding structural aspects of Lebanese Arabic which will be relevant in the discussion of argument doubling. LA is a null subject language as illustrated in (4a-b):

(4) a. kariim ʔakal suuʃi
K. ate.3MS sushi
'Karim ate sushi.'

b. ʔakal suuʃi
ate.3MS sushi
'He ate sushi.'

Furthermore, the word order on a sentential level may be SVO, as in (4a), or VSO, as in (5a). The order VOS is also available (5b); however, it is less common, and a pause is necessary before the subject:

(5) a. ʔakal kariim suuʃi
ate.3MS K. sushi
'Karim ate sushi.'

b. ʔakal suuʃi kariim
ate.3MS sushi K.
'Karim ate sushi.'

3. DISTRIBUTION OF CLITICS

In LA, a pronominal clitic appears with a verb (6a), a noun (6b), or a preposition (6c). In (6a-c), the clitic is in bold (irrelevant details omitted).

(6) a. kariim ʃeef-**o**
K. saw-him
'Karim saw him.'

b. kariim ʃeef kteeb-**o**
K. saw book-his
'Karim saw his book.'

c. kariim raaɦ maʕ-**o**
K. left with-him
'Karim left with him.'

As illustrated in (6a-c), each pronominal clitic appears on the head it is selected by. In this respect, cliticization is a local phenomenon in Lebanese Arabic which does not have a process of 'clitic climbing' such as the one found in various Romance languages (see Kayne (1975, 1984), Rouveret and Vergnaud (1980), and the references mentioned there). The local nature of cliticization in LA will be discussed further in section 6.

A morphological distinction exists for some persons between accusative clitics, dative clitics, and genitive clitics attached to a noun or a preposition. This distinction can be seen, for instance, in examples (7a-d) involving a first person singular clitic:

(7) a. kariim ʃeef-**ne**
K. saw-me
'Karim saw me.'

b. kariim ɦəkee-**le** ɦkeeye
K. told-me story
'Karim told me a story.'

c. kariim ʃeef kteeb-**e**
K. saw book-my
'Karim saw my book.'

d. kariim raaɦ maʕ-**e**
K. went with-me
'Karim went with me.'

4. WHAT CAN BE DOUBLED?

All the clitics discussed in the previous section can be doubled. That is, an accusative clitic (8a), a dative clitic (8b), a genitive clitic attached to a noun (8c), or a preposition (8d), can be doubled. The doubled element and the doubling clitic are in bold in (8).

(8) a. kariim ʃeef-**o** **la-saami**
K. saw-him to-S.
'Karim saw Sami.'

b. kariim ɦəkee-**lo** **la-saami** ɦkeeye
K. told-him to-S. story
'Karim told Sami a story.'

c. kariim ʃeef kteeb-**o** **la-saami**
K. saw book-his to-S.
'Karim saw Sami's book.'

d. kariim raaɦ maʕ-**o** **la-saami**
K. went with-him to-S.
'Karim went with Sami.'

As illustrated in (8a-d), the doubled element appears with the preposition *la* ('to'), thus conforming to Kayne's generalization informally stated in (9) (see Aoun (1981), Jaeggli (1982, 1986), Borer (1983, 1986), Dobrovie-Sorin (1990), and the references mentioned there).

(9) *Kayne's Generalization*
A lexical NP may be doubled by a clitic only if this clitic is preceded by a (prepositional) case assigner. The nominal element following this case assigner is assigned dative case: the pronominal element following it is morphologically dative.[1]

(10) a. kariim ʃeef-**o** **la-ʔəlo**
K. saw-him to-him
'Karim saw him.'

b. kariim ɦəkee-**lo** **la-ʔəlo** ɦkeeye
K. told-him to-him story
'Karim told him a story.'

c. kariim ʃeef kteeb-**o** **la-ʔəlo**
K. saw book-his to-him
'Karim saw his book.'

d. kariim raaɦ maʕ-**o** **la-ʔəlo**
K. went with-him to-him
'Karim went with him.'

In (8a-d) and (10a-d), I gave examples where the complements of a verb, noun, or preposition are doubled. Subjects cannot be doubled as illustrated in (11)-(12):

(11) (*la)-kariim neem
to-K. slept.3MS
'Karim slept.'

(12) neem (*la)-kariim
slept.3MS to-K.
'Karim slept.'

Sentences (11) and (12) indicate that the subject in SV or in VS sentences cannot be doubled. In Exceptional Case-marking constructions, however, the subject can be doubled:

(13) badde-**yee** yruuɦ **la-kariim**[2]
want.1S-him go.3MS to-K.
'I want Karim to go.'

(14) χallayt-**o** yruuɦ **la-kariim**
let.1S-him go.3MS to-K.
'I let Karim go.'

It is worth mentioning that doubling is optional. In all the examples discussed, the clitic can appear by itself. Alternatively, the lexical NP can also appear by itself. In this case, the preposition *la* cannot occur with the non-doubled NP unless this NP is a dative complement as in (15):

(15) kariim ʃeef (*la)-saami (see 8a)
K. saw.3MS (*to)-S.
'Karim saw Sami.'

(16) kariim ɦəke *(la)-saami ɦkeeye (see 8b)
K. told *(to)-S. story
'Karim told Sami a story.'

(17) kariim ʃeef kteeb (*la)-saami (see 8c)[3]
K. saw.3MS book (*to)-S.
'Karim saw Sami's book.'

(18) kariim raaɦ maʕ (*la)-saami (see 8d)
K. went with (*to)-saami
'Karim went with Sami.'

(19) badde kariim yruuɦ (see 13)
want.1S K. go.3MS
'I want Karim to go.'

(20) χalleet kariim yruuɦ (see 14)
let.1S K. go.3MS
'I let Karim go.'

The previous examples illustrate the syntactic positions (complements of V, N, P, and subjects of ECM complements) that can be doubled.

5. INTERPRETATION OF DOUBLED ELEMENTS

In the previous section, we saw what elements can be doubled. In the following sections, I will discuss how to structurally represent doubled elements. First, I address the question of whether doubled elements are interpreted in the same position as non-doubled elements.

5.1. Doubled adnominals

Consider a sentence such as (21):

(21) ʃəft Suurit waladeen (mən hal-wleed)
saw.1S picture boy.dual (of these-the-boys)
'I saw a picture of two (of these) boys.'

Sentence (21) is ambiguous. The quantificational expression *waladeen* ('two boys') may have wide scope (or a distributive reading): 'I saw two pictures, each representing one boy'. The wide scope reading may be highlighted by adding the following paraphrase to the sentence in (20): *wəɦde mənun keenit maχzuu ʔa* ('one of them was torn'). It may also have narrow scope: 'I saw one picture representing two boys' (Fiengo and Higginbotham (1981), May (1985), and Aoun and Hornstein (1985)).

In contrast to sentence (21), sentence (22) in which the whole NP *Suurit waladeen* ('a picture of two boys') is doubled is not ambiguous: the quantificational element has only an internal, narrow scope reading:

(22) **ʃəft-a la-Suurit waladeen (mən hal-wleed)**
saw.1S-it to-picture boy.dual (of these-the-boys)
'I saw a picture of two (of these) boys.'

The contrast between sentences (21) and (22) indicates that a quantificational element cannot be extracted at LF, out of a doubled element, hence, the non-ambiguity of (22). The contrast between the two sentences above may be made clearer with the following pair: since a back (bone) cannot be common to two persons, sentence (23b) is unacceptable.

(23) a.

faɦaSt	Dahr	waladeen	(mən	hal-wleed)
examined.1S	back	boy.dual	(of	these-the-boys)

'I examined the back of two (of these) boys.'

b.*

faɦaSt-o	**la-Dahr**	**waladeen**	**(mən**	**hal-wleed)**
examined.1S	back	boy.dual	(of	these-the-boys)

'I examined the back of two (of these) boys.'

In brief, in a configuration such as (24a), the quantificational element may have a wide and a narrow scope reading, and in a configuration such as (24b), it has only a narrow scope reading:

(24) a. .... V .... [N QP] ......
b. .... V + **clitic** ..... **la-[N QP]** ....

Consider now the following representation:

(24) c. .... V + **clitic** ..... **la-N + clitic la-QP** ....

Representation (24c) contains two doubled elements: the object of the verb and the adnominal complement have been doubled. If the doubled QP in (24c) were in the same position as the non-doubled QP in (24b), we would expect the doubled QP to only have narrow scope. The fact is that in a configuration like (24c), the doubled QP must have wide scope: in sentence (25), which corresponds to (24c), the quantificational element must have wide scope:[4]

(25)

ʃəft-a	**la-Suurit-un**	**la-waladeen**	**(mən**	**hal-wleed)**
saw.1S-it	to-picture-them	to-boy.dual	(of	these-the-boys)

'I saw a picture of two (of these) boys.'

The contrast between (22) and (25) indicates that the doubled adnominal quantifier is not interpreted in the same position as the non-doubled adnominal quantifier: it is in a position higher than the non-doubled element.

For completeness, I provide sentence (26), in which the adnominal quantificational element is doubled. This doubled quantificational element may have wide scope:

(26)

ʃəft	Suurit-**un**	**la-waladeen**	**(mən**	**hal-wleed)**
saw.1S	picture-their	to-boy.dual	(of	these-the-boys)

'I saw a picture of two (of these) boys.'

5.2. Doubled accusatives and doubled obliques

The previous discussion highlighted the fact that a doubled adnominal complement is not interpreted in the position of a non-doubled adnominal complement. A similar argument can be constructed for doubled accusatives, as well as doubled prepositional complements (obliques).

(27) a.* mʕalləmt-[o]$_1$ ʔaaSaSit [kəll walad]$_1$
teacher-his punished.3SF every boy
'His teacher punished every boy.'

b. mʕalləmt-[o]$_1$ ʔaaSaSit-**o** **la-** **[kəll walad]$_1$**
teacher-his punished.3SF-him to every boy
'His teacher punished every boy.'

(28) a.* χabbart mʕalləmt-[o]$_1$ ʕann [kəll walad]$_1$
told.1S teacher-his about every boy
'I told his teacher about every boy'

b. χabbart mʕalləmt-[o]$_1$ ʕann-**o** **la-** **[kəll walad]$_1$**
told.1S teacher-his about-him to every boy
'I told his teacher about every boy'

(29) a.* ʃəft l-mara yalli htammit fi-[i]$_1$ ɦadd [kəll walad]$_1$
saw.1S the-woman that cared.3SF in-him next every boy
'I saw the woman that took care of him next to every boy.'

b. ʃəft l-mara yalli htammit fi-[i]$_1$ ɦadd-**o** **la-** **[kəll walad]$_1$**
saw.1S the-woman that cared.3SF in-him next-him to every boy
'I saw the woman that took care of him next to every boy.'

In (27a), as in (28a) and (29a), the bound reading of the pronoun contained within the DP *m ʕalləmto* ('his teacher'), as well as the pronoun contained within the relative clause *yalli htammit fii* ('that took care of him'), cannot obtain. Assuming the c-command requirement on bound anaphora, the ungrammaticality of (27a-29a) may be attributed to the failure of c-command to hold between the QP *kəll walad* ('every boy') and the pronoun. In contrast, in (27b), (28b), and (29b), the bound reading is available, which indicates that the doubled QP *kəll walad* ('every boy') is able to c-command the pronoun within the DP *m ʕalləmto* ('his teacher') and the relative clause *yalli htammit fii* ('that took care of him'), respectively. The contrast observed between the (a) sentences in (27)-(29) and the (b) sentences leads to the conclusion that doubled accusative complements, as well as doubled prepositional complements are interpreted in a position higher than that of their corresponding non-doubled arguments.[5]

6. DOUBLING, PREDICATION, AND DISJOINT REFERENCE

The question that can now be raised is the following: why can't the doubled element be interpreted in the same position as the non-doubled element? A natural answer would be to say that the complement position is already filled. What are the likely candidates to occur in this position? At this point, it is possible to assume that the complement position is filled by the pronominal clitic, which gets incorporated into the head selecting it ((30)-(31)) (see Kayne (1975)).

(30) X clitic *la* NP

(31) X + clitic_1 t_1 *la* NP

Another possibility would be to assume that the clitic is not generated by movement. Rather, it is coindexed with a non-overt pronominal in the complement position (32) (as argued in Chomsky (1982), Jaeggli (1986), Sportiche (1992), Kayne (1994)):

(32) X + clitic_1 pro_1 *la* NP

According to the analysis in (30)-(31), the clitic itself is a pronominal element and according to the one in (32), the clitic is coindexed with a non-overt pronominal and is not a true argument. There are reasons to favor the analysis in (32) over the one in (30)-(31):

(i) As mentioned earlier, cliticization in Lebanese Arabic has a strictly local character (see Mohammad (1990)): the clitic is attached to the element it is selected by: clitics corresponding to a complement of a noun, a verb, or a preposition get attached to this noun, verb, or preposition, respectively (33a-c). There is no process of 'clitic climbing' in this language as opposed, for instance, to Romance languages such as French, Spanish, and Italian, where the clitic does not necessarily appear attached to the head that selects it (see Kayne (1975, 1984), Rouveret and Vergnaud (1980)):

(33) a. V + clitic b. N + clitic c. P + clitic

(ii) It is also the case that a clitic in LA may co-occur with an overt non-doubled argument. This is illustrated in (34): as argued in Benmamoun (1992) for Arabic and Shlonsky (1991) for Hebrew, the nominal element in (35a) occurs in the Specifier of the quantifier and enters into Spec-Head agreement with this quantifier (see also Sportiche (1992) and Koopman (1991)). This accounts for the occurrence of the agreeing clitic in (35a). If clitics were arguments, we would not expect them to co-occur with another argument; otherwise, we would have a violation of the Thematic Theory (see Chomsky (1991, chapter 2)): two arguments would share a unique thematic role.

(34) ʃəft kəll l-wleed
saw.1S all the-boys
'I saw all the boys.'

(35) a. ʃəft [$_{QP}$ [$_{Spec}$ l-wleed] [$_{Q'}$ kəll-un]]
saw.1S the-boys all-them
'I saw all the boys.'

(iii) Finally, two clitics may correspond to a unique argument, as illustrated in (35b) where the clitic attached to the verb and the one attached to the noun correspond to the same argument (see Benmamoun (1992) for an analysis of these constructions):[6]

(35) b. ʃəft-un [$_{QP}$ kəll-un]]
saw.1S-them all-them
'I saw all of them.'

Representation (32) captures the fact that the doubled element does not occur in the same position as the non-doubled complement. There is a pronominal element, the non-overt *pro*,[7] obligatorily coindexed with the doubled element, which is generated in the argument position.

6.1. Doubling and predication

Since doubled elements are not generated in the same position as non-doubled elements, one may wonder how they get interpreted. There are two means for an argument to be interpreted: an argument may be interpreted by virtue of receiving a thematic role from a head X^0 that governs it or by virtue of being coindexed with a predicate. The first situation obtains with complements which get their thematic roles directly from the head selecting them. The second situation obtains with subjects which get interpreted with respect to a predicate. As in Williams (1980), the predicate may be a simple predicate; in this case, the subject is the external argument of the head of the predicate and appears outside the maximal projection of the head, as in (36). In (36), the subject is underlined:

(36) <u>John</u> [$_{VP}$ ate my lunch]

The predicate may also be complex. In Williams (1980), it is indicated that sentences function as complex predicates when they contain a 'predicate variable' linked to the subject of predication. According to Williams (1980), the predicate variable is the open position in the sentence which makes it a one-place predicate. This predicate variable may be a pronominal (PRO or *pro*) (see Williams (1980), Rothstein (1983), Browning (1987), and Iatridou (1991)). The structural relation that must hold between the predicate and its subject is mutual c-command.

Returning to doubling in LA, the doubled element in LA is not in a complement position; it has to be interpreted with respect to a predicate. The forced co-indexing between the doubled element and the clitic-*pro* complex may be viewed as resulting from the predication relation holding between the doubled element and the phrase it is attached to. In other words, I am suggesting that the doubled element has the role of a subject of predication, the phrase it is attached to is a predicate and the *pro* argument provides the open position, or the predicate variable.[8]

6.2. Doubling and disjoint reference

Assuming that the doubled element is obligatorily coindexed with a null pronominal element leads us to expect a disjointness effect to exist between the null pronominal and the doubled element. I will argue here that this is indeed the case.

In Lebanese Arabic, disjointness holds between a subject and an object, or between a subject and certain prepositional complements. In sentences (37a-c), the pronominal object and the subject must be disjoint in reference:

(37) a. kariim ʃeef-o
K. saw.3SM-him
'Karim saw him.'

b. kariim ɦəkee-lo ɦkeeye
K. told.3SM-him story
'Karim told him a story.'

c. kariim ɦəke maʕ-o
K. spoke.3SM with-him
'Karim spoke with him.'

In these sentences, an anaphor bound to the subject can occur in object position:

(38) a. kariim ʃeef ɦaalo
K. saw.3SM himself
'Karim saw himself.'

b. kariim ɦəkee la-ɦaalo ɦkeeye
K. told.3SM to-himself story
'Karim told himself a story.'

c. kariim ɦəke maʕ-ɦaalo
K. spoke.3SM with-himself
'Karim spoke with himself.'

It goes without saying that an object or a prepositional complement can be coreferential with another noun phrase when a subject intervenes between them. In (39a-c), the matrix subject and the embedded pronoun can be coreferential:

(39) a. kariim ʔaal ʔənno zeena ʃeefət-o
K. said.3SM that Z. saw.3SF-him
'Karim said that Zeina saw him.'

b. kariim ʔaal ʔənno zeena ɦəkyit-lo ɦkeeye
K. said.3SM that Z. told.3SF-him story
'Karim said that Zeina told him a story.'

c. kariim ʔaal ʔənno zeena ɦəkyit maʕ-o
K. said.3SM that Z. spoke.3SF with-him
'Karim said that Zeina spoke with him.'

On the other hand, disjointness does not hold between an adnominal complement and a subject:

(40) kariim ʃeef Suurt-o
K. saw.3SM picture-his
'Karim saw his picture.'

An anaphor cannot occur in an adnominal complement position:

(41) * kariim ʃeef Suurit ɦaalo
K. saw.3SM picture himself
'Karim saw a picture of himself.'

As a descriptive generalization, there are at least two domains in which a pronoun has to be free: the minimal phrase containing a subject or the minimal noun phrase in which it is contained. We will refer to these domains as Complete Functional Complexes or CFCs (see Chomsky (1986)).

Returning to doubling in LA, I would like to argue that a disjointness effect regulates the relation between a doubled element and the doubling clitic-*pro*. We have already seen that noun phrases, and minimal phrases containing a subject constitute CFCs. Prepositions such as *ɦadd* ('next to') or *ʔədeem* ('in front of') also constitute CFCs. Thus, the null pronominal complement of *ɦadd* ('next to') or *ʔədeem* ('in front of') in (42) below can be coreferent with the subject *Karim*. Furthermore, as (42a-b) show, an anaphor related to the subject may not occur in the complement position of such prepositions. Henceforth, I refer to prepositional phrases which constitute a CFC as Complex PPs.

(42) a. kariim$_1$ ʔaʕʕad marwaan ɦadd-[o]$_1$ / *ɦaalo$_1$
K. sat.3SM M. next-him himself
'Karim sat Marwan next to him'

b. kariim$_1$ ʃeef ɦayye ʔəddeem-[o]$_1$ / *ɦaalo$_1$
K. saw.3SM snake in-front-him himself
'Karim saw a snake in front of him.'

However, all PPs do not constitute CFCs (I will henceforth refer to such prepositional phrases as Simplex PPs), as evidenced by the ungrammaticality of the following sentences:

(43) a.* kariim$_1$ ɦəke ʕann-[o]$_1$
K. talked.3SM about-him
'Karim talked about him.'

b.* kariim$_1$ ɦəke maʕ-[o]$_1$
K. talked.3SM with-him
'Karim talked with him.'

In contrast with (42a-b) above, the pronominal complement of a preposition such as *ʕan* ('about') or *maʕ* ('with') cannot be coreferent with the subject in (43a-b).

Assuming that doubled elements enter into disjointness with the clitic-*pro* complex, they have to occur outside the CFC containing this clitic-*pro* complex. Otherwise, a non-well-formed predication results (44), and the doubled element fails to be interpreted:

(44) * [$_{CFC}$ *pro$_2$* .... doubled element$_1$]

With this in mind, consider the following sentences:

(45) ʃəft ʔəmm-**o** **la-kariim** mbeeriɦ
saw.1S mother-his to-K. yesterday
'I saw Karim's mother yesterday.'

In sentence (45), the doubled element, the doubling clitic, and the element this clitic is attached to (the host), can be Clitic-Left-Dislocated: they can appear at the beginning of the clause and be coindexed with a clitic, as in (46) (see Cinque (1990), Iatridou (1991), Demirdache (1991)):

(46) ʔəmm-**o** **la-kariim** ʃəft-a mbeeriɦ
mother-his to-K. saw.1S-her yesterday
'Karim's mother, I saw her yesterday.'

They can be also pied-piped by wh-movement as in (47). Under standard assumptions according to which extraction processes affect single constituents, it is safe to conclude that in (45)-(47), the clitic, its host, and the doubled element, form a single constituent:

(47) ʔəmm-**o** **la-ʔayya** **walad** ʃəft mbeeriɦ
mother-his to-which boy saw.2MS yesterday
'Which boy's mother did you see yesterday?'

Sentences (48)-(49) lead to the same conclusion:

(48) a. ɦadd-**o** **la-kariim** ʃəft ɦayye mbeeriɦ
near-him to-K. saw.2MS snake yesterday
'Near Karim you saw a snake yesterday.'

b. ʔəddem-**o** **la-kariim** weeʔif rəʒʒeel Tawiil
front-him to-K. standing.M man tall
'A tall man is standing in front of Karim.'

(49) a. ɦadd-**o** **la-ʔayya** **walad** ʃəft ɦayye mbeeriɦ
near-him to-which boy saw.2MS snake yesterday
'Near which boy did you see a snake yesterday?'

b. ʔəddem-**o** **la-ʔayya** **walad** weeʔif rəʒʒeel Tawiil
front-him to-which boy standing.M man tall
'In front of which boy is a tall man standing?'

In (48)-(49), the preposition, the clitic and the doubled element may be treated as a single constituent, and fronted by wh-movement. We have already seen that noun phrases and PPs headed by *ɦadd* ('next to') or *ʔəddeem* ('in front of') form CFCs. The grammaticality of (46)-(47) and (48)-(49) goes to show that the doubled element can be attached to, and form a constituent with these CFCs.

Consider now the following paradigm:

(50) a.* ʕann-**o** / maʕ-**o** **la-kariim** ɦkiit
about-him with-him to-K. spoke.2SM
'About/With Karim you spoke.'

b.* ʕann-**o** / maʕ-**o** **la-ʔayya** **walad** ɦkiit
about-him with-him to-which boy spoke.2SM
'About/With which boy did you speak?'

Since PPs headed by *ʕan* ('about') or *maʕ* ('with') do not form a CFC, the doubled element cannot attached to these phrases and be topicalized or extracted by wh-movement. Hence the ungrammaticality of (50).

Note that if the prepositional clitic in (50) is further embedded within a DP, as in (51), (wh-)extraction becomes possible again:

(51) a. ʕann / maʕ ʔəmm-o **la-kariim** ɦkiit
about with mother-his to-K. spoke.2SM
'About/With Karim's mother you spoke.'

b. ʕann / maʕ ʔəmm-o **la-ʔayya** **walad** ɦkiit
about with mother-his to-which boy spoke.2SM
'About/With which boy's mother did you speak?'

The sentences in (51) clearly pattern with those in (48)-(49): in these sentences, the doubled element is attached to an opaque domain, i.e. the noun phrase which contains the doubling clitic-*pro*.

Summarizing the results in this section, doubled elements are interpreted as subjects, with respect to a complex predicate. This predicate contains a null pronominal which is generated in argument position and is coindexed with the doubled element. As a consequence of the disjointness requirement holding between the null pronoun in argument position and the doubled element, this predicate must be a CFC. In LA, nominal phrases, prepositional phrases headed by *ɦadd* ('next to') or *ʔədeem* ('in front of'), and minimal phrases containing subjects were identified as CFCs.

7. WHAT CONSTITUTES A COMPLEX PREDICATE?

In what follows, I will provide evidence that the complex predicate with which doubled arguments enter into a predication relation is in fact the **minimal** CFC containing the predicate variable. The evidence is based on the binding properties of doubled adnominals and doubled obliques attached to complex PPs.

7.1. Doubled adnominals

We saw that doubled adnominals are in a position higher than that of non-doubled adnominals (section 5.1.) and that they can form a constituent with a DP containing the doubling clitic.

With this in mind, consider the following sentences:

(52) a. [kəll bənt]$_1$ ʕatit l-mudiir ʕalemeeto **la -χayy-[a]$_1$**
every girl give.3SF the-principal grades-his to-brother-her
'Every girl gave the principal her brother's grades.'

b. ʕatiit [kəll bənt]$_1$ l-mudiir ʕalemeeto **la -χayy-[a]$_1$**
gave.3SF every girl the-principal grades-his to-brother-her
'Every girl gave the principal her brother's grades.'

c. l-mudiir ʕaTa [kəll bənt]$_1$ ʕalemeeto **la-χayy-[a]$_1$**
the-principal gave.3SM every girl grades-his to-brother-her
'The principal gave every girl her brother's grades.'

In all the above sentences, the bound reading of the pronoun within the doubled adnominal is available. (52a-b) show that the preverbal subject, as well as the post-verbal subject can c-command this pronoun, and (52c) illustrates the same point for the direct object. The paradigm in (52) indicates that the pronoun contained within the doubled adnominal is in a position which can be c-commanded by the arguments of the clause within which the doubled element occurs. The paradigm in (52) may be accounted for if doubled adnominals **can** form a constituent with the minimal DP containing the doubling clitic, i.e. the DP *ʕalemeeto* ('his grades') in (52a-c). As such, they will be lower than the other arguments (subject, object) of the clause.

The above conclusion can be strengthened. It is possible to show that doubled adnominals **must** form a constituent with the minimal DP containing the doubling clitic. Consider the following paradigm:

(53) a.* farʒeet ʕalemeet-[o]$_1$ la-ʔəmmo **la-[kəll walad]$_1$**
showed.1S grades-his to-mother-his to- every boy
'I showed his grades to every boy's mother.'

b.* ʕaTit mʕalləmt-[o]$_1$ l-mudiira ʕalemeeto **la-[kəll walad]$_1$**
gave.3SF teacher-his the-principal grades-his to-every boy
'The teacher gave the principal every boy's grades.'

c.* mʕalləmt-[o]$_1$ ʕaTit l-mudiira ʕalemeeto **la-[kəll walad]$_1$**
teacher-his gave.3SF the-principal grades-his to-every boy
'The teacher gave the principal every boy's grades.'

(53a-c) are all ungrammatical under the intended bound reading. In (53a), the pronoun contained within the direct object *ʕalemeeto* ('his grades') cannot be bound by the doubled adnominal QP *kəll walad* ('every boy'). In (53b) and (53c), the pronoun contained within the post-verbal subject and the preverbal subject respectively, cannot be bound by this doubled adnominal either.

Consider now the following sentences:

(54) a. farʒeet [ʕalemeeto **la-[kəll walad]$_1$**] la-ʔəmm-[o]$_1$
showed.1S grades-his to- every boy to-mother-his
'I showed every boy's grades to his mother.'

b.

ʕaTit	mʕalləmto	**la-[kəll walad]$_1$**	l-mudiira	ʕalemeet-[o]$_1$
gave.3SF	teacher-his	to-every boy	the-principal	grades-his

'Every boy's teacher gave the principal his grades.'

c.

mʕalləmto	**la-[kəll walad]$_1$**	ʕaTit	l-mudiira	ʕalemeet-[o]$_1$
teacher-his	to-every boy	gave.3SF	the-principal	grades-his

'Every boy's teacher gave the principal his grades.'

Sentences (54a-c) minimally contrast with those in (53a-c). Clearly, sentences (53a-c) could not have been derived from sentences (54a-c), otherwise we would expect sentences (53a-c) to display the same binding possibilities as the ones displayed in sentences (54a-c).[9]

The unavailability of a bound reading in (53a-c) illustrates a standard weak crossover violation: we saw in section 5.1. that the doubled adnominal can have scope outside the DP it is attached to. At LF, the variable of the QP left by QR will be coindexed with a pronoun to its left. In (54a-c), no weak crossover configuration arises: the variable of the doubled adnominal precedes the bound pronoun.

Finally, consider the following sentences:

(55) a.

[**farʒayto**	ʕalemeet-[o]$_1$]	**la-[kəll**	**walad]$_1$**
showed.1S-him	grades-his	to- every	boy

'I showed his grades to every boy.'

b.

[**ʕaTito**	mʕalləmt-[o]$_1$	l-ʕalemeet]	**la-[kəll walad]$_1$**
gave.3SF-him	teacher-his	the-grades	to-every boy

'His teacher gave every boy his grades.'

c.

[mʕalləmt-[o]$_1$	**ʕaTito**	l-ʕalemeet]	**la-[kəll walad]$_1$**
teacher-his	gave.3SF-him	the-grades	to-every boy

'His teacher gave every boy his grades.'

In sentences (55a-c), the doubled element can bind the adnominal pronoun contained within the direct object in (55a), or the subject in (55b-c).[10] In (55a-c), the clitic attached to the verb may be taken to be the doubling clitic. The CFC containing this doubling clitic is the whole clause. This CFC and the doubled element are in a predication relation: the doubled element c-commands the whole complex predicate, and, as such, also c-commands the adnominal clitic contained within the complex predicate.[11]

The fact that the doubled QP can bind the pronoun in (55a-c) but not in (53a-c) indicates that a doubled element has to be generated with the minimal CFC containing the doubling clitic. In (53a-c), the minimal CFC containing the doubling clitic is the DP and in (55a-c), it is the clause.

Summing up, the interpretive facts involving doubled adnominals illustrate the following generalizations:

(56) a. Doubled adnominals are attached to the minimal CFC, i.e. the DP which contains the null argument coindexed with the doubling clitic.

b. Doubled adnominals cannot bind a pronoun contained within a preceding argument of the minimal clause in which they appear.

7.2. Doubled obliques attached to Complex PPs

Turning to Complex PPs headed by prepositions such as *ɦadd* ('next to') or *ʔədeem* ('in front of'), if, as suggested earlier, these Complex PPs are also CFCs, then we expect a doubled oblique related to these Complex PPs to have a behavior similar to a doubled adnominal related to a DP. This expectation is borne out, as demonstrated by the following sentences:

(57)	a.*	ʔəmm-[o]$_1$	ʃeefit	ɦayye	ɦaddo	**la-[kəll walad]$_1$**
		mother-his	saw.3SF	snake	next-him	to-every boy
		'His mother saw a snake next to every boy.'				
	b.*	ʃeefit	ʔəmm-[o]$_1$	ɦayye	ɦaddo	**la [kəll walad]$_1$**
		saw.3SF	mother-his	snake	next-him	to-every boy
		'His mother saw a snake next to every boy.'				

Just like doubled adnominals, the doubled complement of *ɦadd* ('next to') cannot bind a pronoun contained within a subject in the clause in which this doubled element occurs. The paradigm in (57), like the paradigm in (53), is accounted for if the doubled complement of the preposition *ɦadd* ('next to') is attached to the minimal CFC -- the Complex PP headed by *ɦadd* ('next to') -- containing the doubling clitic.

Furthermore, as was the case for doubled adnominals (52), the doubled complement of *ɦadd* ('next to') can be bound by the preverbal and post-verbal subjects (58a-b).

(58)	a.	[kəll bənt]$_1$	ʃeefit	ɦayye	ɦaddo	**la-χayy-[a]$_1$**
		every girl	saw.3SF	snake	next-him	to-brother-her
		'Every girl saw a snake next to her brother.'				
	b.	ʃeefit	[kəll bənt]$_1$	ɦayye	ɦaddo	**la-χayy-[a]$_1$**
		saw.3SF	every girl	snake	next-him	to-brother-her
		'Every girl saw a snake next to her brother.'				

Finally, if the doubled prepositional complement of *ɦadd* ('next to') is attached to the Complex PP headed by this preposition, then we expect the doubled element to be able to bind a pronoun contained within the 'subject' of that PP. This is indeed the case, as illustrated in (59):[12]

(59) zeena ʃeefit ʔəmm-[o]$_1$ ɦaddo **la-[kəll walad]$_1$**
Z. saw.3SF mother-his next-him to-every boy
'Zeina saw his mother next to every boy.'

The contrast between sentences (56a-b) and sentence (59) can be accounted for if doubled elements related to an oblique clitic are attached to the minimal CFC -- in this case, the Complex PP -- which contains the oblique clitic. To sum up, the interpretive facts involving doubled complements of prepositions such as *ɦadd* ('next to') are parallel to those involving doubled adnominals:

(60) a. Doubled obliques associated with complex PPs are attached to the minimal CFC -- the Complex PP -- which contains the null argument coindexed with the doubling clitic.

b. Doubled obliques attached to complex PPs cannot bind a pronoun contained within a preceding argument of the minimal clause in which they appear (modulo the case of the 'subject' of complex PPs (see footnote 12)).

8. GENERATION OF DOUBLED ELEMENTS

In the previous section, we suggested that doubled elements function as subjects that enter into a predication relation with the minimal CFC containing the predicate variable:

(61) [[$_{CFC}$ .... *pro$_1$* ....] doubled element$_1$]

Representation (61), which, incidentally, violates the Linear Correspondence Axiom (LCA) of Kayne (1994), raises the following question: Subjects in LA are usually to the left of their predicate (see Aoun, Benmamoun, and Sportiche (1994)). If indeed the doubled element functions as the subject of a complex predicate, why is it generated to the right of its predicate?

A possible answer to the question raised above, is to assume that representation (61) is a derived representation. An obvious candidate for the representation underlying (61) is the one in (62):

(62) [doubled element$_1$ [$_{CFC}$ .... *pro$_1$* ....]]

We now discuss some evidence suggesting that (62) may indeed be the representation underlying (61).

Notice first, that a representation such as (61) cannot be adopted as the representation for all doubled accusatives. That is, it cannot be suggested that doubled accusatives always appear in a right c-commanding position with respect to the CFC to which they are related. There are constructions in which doubled accusatives are clearly lower than the constituents to the right of which they appear. One such construction is exemplified by doubled accusatives in Exceptional Case Marking contexts, such as (63) (see Sportiche (1992)):

(63) a. zeena χalləto **la-kariim** yruuħ ʕal-madrase
Z. let.3SF-him to-K. go.3SM to-the-school
'Zeina let Karim go to school.'

b. ʔəmm-[o]$_1$ χalləto **la-[kəll walad]$_1$** yruuħ ʕal-madrase
mother-his let.3SF-him to-every boy go.3SM to-the-school
'His mother let every boy go to school.'

In (63a), the doubled accusative occurs between the matrix verb and the embedded clause. Nevertheless, the doubled accusative can bind a pronoun contained within the preverbal subject, as in (63b). In brief, sentences such as (63) raise a problem for an analysis that assumes that (61) is the non-derived representation of doubled accusatives.

No problem arises, however, in case representation (62) is assumed to be the underlying representation for doubled accusatives. From this underlying representation, two possible types of derivation are available. In the first type (type a), the whole CFC is raised to the left of the doubled element. In the second type (type b), individual elements within the CFC are raised to the left of the doubled element.[13]

To illustrate, consider doubled accusatives in ECM contexts. The underlying representation for these constructions will be as in (64):

(64) [doubled element$_1$ [$_{CFC}$ ..... subject .... V ... [$_{embedded\ clause}$ ....]]]

The whole CFC in (64) could be raised to the left of the doubled element, thus generating (61). Alternatively, the subject and the verb may be raised by distinct operations to the left of the doubled element, thus generating (65a) or (65b):

(65) a. subject$_1$ ... V$_2$ ......doubled element$_1$ [$_{CFC}$ ..... t_1 ... t_2 .... [$_{embedded\ clause}$ ....]]

b. V$_2$ ...... subject$_1$ ... doubled element$_1$ [$_{CFC}$ ..... t_1 ... t_2 ....[$_{embedded\ clause}$ ....]]

The specific examples (66a-b) and (63a-b) correspond to representations (61) and (65a), respectively:

(66) a. zeena **χalləto** yruuħ ʕal-madrase **la-kariim**[14]
Z. let.3SF-him go.3SM to-the-school to-K.
'Zeina let Karim go to school.'

b. ʔəmm-[o]$_1$ **χalləto** yruuħ ʕal-madrase **la-[kəll walad]$_1$**
mother-his let.3SF-him go.3SM to-the-school to-every boy
'His mother let every boy go to school.'

Certain binding facts follow in case the analysis outlined in (61)-(65) is adopted. For instance, the surprising fact that the pronoun contained within the preverbal subject can be bound by the doubled element in ECM contexts can now be treated in terms of reconstruction: the preverbal subject reconstructs to its trace position -- below the doubled element. As such, the doubled element can bind a pronoun contained within this subject.

More generally, the above analysis predicts the existence of the following generalization:

(67) doubled accusatives, doubled datives and doubled obliques attached to simplex PPs (henceforth, doubled obliques) can bind a pronoun contained within a preceding or following argument of the minimal clause in which they appear.

According to the analysis outlined in (63)-(67), a doubled element is higher than the arguments contained within the minimal CFC that functions as its complex predicate. Any argument of this complex predicate which precedes the doubled element must have been raised from a position below this doubled element. As such, this argument can reconstruct to this lower position, and a pronoun contained within this reconstructed argument can be bound by the doubled element. In a sentence like (63b), the subject can reconstruct to its trace position which is c-commanded by the doubled element (see 65a). In a sentence such as (66b), the whole CFC can reconstruct to its underlying position which is c-commanded by the doubled element (see (61)-(62)).

The following examples provide further illustration of the empirical validity of the generalization in (67):

doubled accusatives

(68) a. **χabbarta** **la-kəll bənt$_1$** ʕan ʔəmm-[a]$_1$
told.1S-her to-every girl about mother-her
'I told every girl about her mother.'

b. **χabbarta** ʕan ʔəmm-[a]$_1$ **la-kəll bənt$_1$**
told.1S-her about mother-her to-every girl
'I told every girl about her mother.'

doubled datives

(69) a. farʒayto **la-kəll walad$_1$** ʕalemeet-[o]$_1$
showed.1S-him to-every boy grades-his
'I showed every boy his grades.'

b. farʒayto ʕalemeet-[o]$_1$ **la-kəll walad$_1$**
showed.1S-him grades-his to-every boy
'I showed every boy his grades.'

doubled obliques

(70) a. χabbart l-mara yalli htammit fiy-[a]$_1$ ʕanna **la-kəll bənt$_1$**
told.1S the-woman that cared.3SF in-her about-her to-every girl
'I told the woman that took care of her about every girl.'

b. χabbart ʕanna **la-kəll bənt$_1$** l-mara yalli htammit fiy-[a]$_1$
told.1S about-her to-every girl the-woman that cared.3SF in-her
'I told about every girl the woman that took care of her .'

9. OTHER ACCOUNTS OF DOUBLING

We are now in a position to briefly discuss various accounts of doubled elements put forward in the relevant literature. These accounts essentially can be regrouped into two main classes. The first assumes that doubled elements are generated in the same argument position in which standard complements are generated (see Jaeggli (1982), Borer (1983), among many others). An interesting variant assumes that doubled elements are generated in the regular complement position (DP*) and are then raised to the Specifier position (DP^) of the clitic which is treated as a head (see Franco (1991), Sportiche (1992)). In this approach, clitic doubled constructions differ from non-doubled constructions in that DP* is overt in the first ones, and covert in the second ones (see Sportiche (1992)):

(71) [DP^ [[clitic] .... DP*]]

The second assumes that doubled elements occupy the position of right-dislocated phrases (Hurtado (1984)). For the sake of completeness, let me mention that a hybrid proposal is put forward in Aoun (1981), where it is argued that, cross-linguistically, some doubled elements are to be treated as arguments, and some are not, even though the exact position of non-argument doubled elements is not made explicit. In this paper, I am not in a position to discuss the cross-linguistic adequacy of each account. The behavior of doubled elements across various languages is not uniform. For instance, a reflexive anaphor can be doubled in various Romance languages (see the above-mentioned references). This is not the case in LA, as we mentioned earlier. It may very well be the case that the various proposals, ultimately, will turn out to be necessary to account for the differences between doubled elements across languages.

My purpose is to review the various proposals in the light of the working of doubled elements in LA. I will investigate the adequacy of the various proposals for LA. I will discuss three alternative analyses for doubled elements in LA:

(72) (i) The doubled element in LA is generated in argument position (DP* in (71)) and it stays in this position in Syntax.

(ii) The doubled element in LA is generated in argument position (DP* in (71)) and then is overtly moved to the Specifier position of the clitic (DP^ in (71)). This possibility is the one explicitly assumed in Sportiche (1992) for LA.

(iii) The doubled element is directly generated in the Specifier of the clitic (DP^ in (71)).

The major problem all three analyses face in LA is the problem of disjointness between the doubled element and the pronominal clitic-*pro* sequence. As mentioned in the previous section, the doubled element can only attach to the minimal complete functional complex dominating the clitic. This accounted for the fact that doubled elements can form a constituent and be pied-piped only with phrases that form an opaque domain. In particular, this accounted for the contrast between sentences (50a-b) and (48a)-(49a), repeated below:

(50) a.* ʕann-o **la-kariim** ɦkiit
about-him to-K. spoke.2SM
'About Karim, you spoke.'

b.* ʕann-o **la-ʔayya walad** ɦkiit
about-him to-which boy spoke.2SM
'About which boy did you speak?'

(48) a. ɦadd-o **la-kariim** ʃəft ɦayye mbeeriɦ
near-him to-K. saw.2MS snake yesterday
'Near Karim you saw a snake yesterday.'

(49) a. ɦadd-o **la-ʔayya walad** ʃəft ɦayye mbeeriɦ
near-him to-which boy saw.2MS snake yesterday
'Near which boy did you see a snake yesterday?'

Such contrasts are problematic for analyses (72i-iii), which assume that doubled arguments are in the regular complement position. The problem is compounded by the fact that a regular non-doubled complement corresponding to (50a-b) forms a constituent with the preposition and can be pied-piped along with the preposition as illustrated in (73a-b). It is not clear how to account for the difference between (50a-b) and (73a-b) in case any of the analyses in (69) is adopted: why is it possible to pied-pipe a doubled element with the prepositional phrase in (48a)-(49a), but not with the one in (50)?

(73) a. ʕann kariim ɦkiit
about K. spoke.2SM
'About Karim, you spoke.'

b. ʕann ʔayya walad ɦkiit
about which boy spoke.2SM
'About which boy did you speak?'

10. CONCLUSION

Let us recapitulate the main features of the analysis advocated so far:

(74) **Analysis**

a. The doubled element is to be analyzed as a subject in a predication relation holding between the doubled element and the minimal complete functional complex containing the predicate variable, i.e. the clitic-*pro*.

b. The doubled element, like all subjects in LA, is generated to the left of the complex predicate with which it is construed.

c. A standard disjointness effect exists between the doubled element and the pronominal predicate variable coindexed with it.

This analysis accounts for the following descriptive generalizations:

(75) **Generalizations**

a. The doubled element is not generated in the argument position selected by the head to which the clitic is attached (see sections 5.1. and 5.2.).

b. The doubled element is generated in a position higher than this argument position (see sections 5.1. and 5.2.).

c. The doubled element can form a constituent with a DP or a complex PP but not with a simplex PP (see section 5.2.).

d. Doubled adnominals cannot bind a pronoun contained within a preceding argument of the minimal clause in which they appear (see section 7.1.).

e. Doubled obliques attached to complex PPs cannot bind a pronoun contained within a preceding argument of the minimal clause in which they appear (see section 7.2.).

f. Doubled accusatives, doubled datives, and doubled obliques can bind a pronoun contained within a preceding or following argument of the complex predicate they are construed with.

As stated in the introduction, these constructions are of interest in that they open a window on the formation of argument structures: they allow us to better understand how some argument relations are formed and how they get to be syntactically projected. Doubling in LA is used to form new functional structures. It is to be viewed as a means to create non-standard argument relations, to superimpose new functional structures on conventional ones.[15]

NOTES

* A version of this paper was first presented at USC and UCLA in the Spring of 1991 and appeared as Aoun (1993). I wish to thank the audiences there. I also benefited from extensive discussions with A. Benmamoun, N. Hornstein, A. Li, J. Ouhalla, M. Mohammad, D. Sportiche, P. Schneider-Zioga, J.-R. Vergnaud, A. Weinberg, and M.-L. Zubizarreta. I wish to thank Z. El-Imad, T. Kadri, J. Kikano-Aoun, T. Attiyeh who were patient and gracious informants. Special thanks to Lina Choueiri for her help and input.

[1] The pronominal element following the marker *la* is a dative pronoun that does not seem to be cliticized onto *la*, as evidenced by the fact that, contrary to other clitics, it can be conjoined:

(i) a.* kariim ɦəkee-**lo** w **la** ɦkeeye
K. told-him and her story
'Karim told him and her a story.'

b. kariim ɦəkee-**lun** **laʔəlo** w **ʔəla** ɦkeeye
K. told-them to-him and to-her story
'Karim told him and her a story.'

In LA, clitics may attach to prepositions (see section 3). The fact that the pronominal elements cannot cliticize to the doubling marker *la* suggests that this marker is not to be analyzed as a preposition. It may be considered as a non-prepositional case-marker. J. Ouhalla (personal communication) suggests that this marker may be considered as a focus marker akin to the focus marker *la* found in Standard Arabic.

[2] In a sequence of two clitics, the second clitic appears with *y* in LA:

(i) ʕTayt-**o-yee**
gave.1S-it-him
'I gave it to him.'

[3] (17) is acceptable with the reading where *la-saami* is not the complement of book, but rather a benefactive: 'Karim saw a book for Sami.'

[4] Once again, the wide scope reading may be highlighted by the acceptability of the following sentence (see 23a-b):

(i) ?fahaSt-o **la-Dahrun** **la-waladeen (mən** **hal-wleed)**
examined.1S-it to-back-their to-boy.dual (of these-the-boys)
'I examined the back of two (of these) boys.'

There is another more natural form for inalienables. It is given in (ii):

(ii) fahaStəllun Dahrun **la-waladeen** **(mən** **hal-wleed)**
examined.1S-to them back-their to-boy.dual (of these-the-boys)
'I examined the back of two (of these) boys.'

[5] It cannot be said that in (27b-28b), the null pronominal coindexed with the clitic attached to the verb *ʔaaSaSit* ('punished'), or the prepositions *ʕan* ('about') and *ħadd* ('next to'), respectively, c-commands the pronoun within the DP *m ʕalləmto* ('his teacher') or within the relative clause *yalli htammit fii* ('that took care of him'). If this were the case, we would expect the following sentences to be ungrammatical, as they would violate principle C of the binding theory. This is contrary to fact:

(i) a. mʕallmit kariim$_1$ ʔaaSaSət-[o]$_1$
teacher K. punished.3SF-him
'Karim's teacher punished him.'

b. χabbart mʕallmit kariim$_1$ ʕann-[o]$_1$
told.1S teacher kariim about-him
'I told Karim's teacher about him.'

c. ʃəft l-mara yalli htammit bi kariim$_1$ ħadd-[o]$_1$
saw.1S the-woman that cared.3SF in K. next-him
'I saw the woman that took care of Karim next to him.'

[6] More generally, the distribution of clitics in LA may be accounted for if clitics are heads that enter into a checking relation with a nominal element (Benmamoun (1992)). In (35a), for instance, the clitic enters into a checking relation with the DP *l-wleed* ('the boys'). In (35b), the clitic attached to the quantifier and the one attached to the verb enters into a checking relation with a non-overt pronominal element, as argued for in Benmamoun (1992) for Standard Arabic and Moroccan Arabic.

[7] This non-overt pronominal may end up in the Specifier of the clitic if the clitic is analyzed as a head whose Specifier gets filled with the element it agrees with as in Chomsky (1991), Franco (1991), and Sportiche (1992) (see also footnote 6). In Browning (1987), it is argued that non-overt operators are pure pronominals, *pro*. Even though her analysis does not mean that every *pro* is to be characterized as a non-overt operator, it raises the possibility of treating the *pro* coindexed with the clitic as a non-overt operator. In this paper, further discussion of this possibility will not be entertained. Some relevant considerations may be found in Schneider-Zioga (1993).

[8] Our analysis of doubling is parallel to the analysis of Modern Greek clitic-left-dislocation in Iatridou (1991), which is accounted for in terms of predication. In Schneider-Zioga (1993), it is argued that the disjointness analysis we put forward accounts for the distribution of clitics as well as the behavior of clitic doubling in Modern Greek.

The relation between the doubled element and the predicate it is related to is reminiscent of that found in complex adjectival constructions such as the following:

(i) John is easy [$_{CP}$ Op$_1$ [PRO to please e$_1$]]

In both constructions, the argument (the subject 'John' in the English adjectival construction and the doubled element in LA) is not in a thematic position and is interpreted with respect to a complex predicate. In Chomsky (1992), complex adjectival constructions are viewed as involving generalized transformations. Such an approach readily extends to doubled constructions in LA.

[9] Alternatively, even if we were to assume that sentences (53a-c) are derived from sentences (54a-c), we would have to conclude that the doubled element in (53a-c) cannot be interpreted in the original position from which it was extracted.

[10] The fact that sentences (55a-c) do not display weak crossover effects is accounted for in section 8. There it is argued that, at LF, the whole CFC to the left of the doubled element reconstructs to a position lower than the doubled element. After reconstruction, the variable of the doubled QP will precede the bound pronoun in (55a-c).

[11] Note that in (55a), the pronoun within the DP object *ʕalemeeto* ('his grades') may be taken to be the predicate variable coindexed with the doubling clitic *-o*. In that case, the minimal CFC containing the doubling clitic is the DP *ʕalemeeto* ('his grades'). The doubled argument would then be attached to and interpreted with respect to this DP. However, under this derivation the QP is not able to bind the dative pronominal in (55a). In other words, (55a) would pattern with (53a).

[12] We are assuming that the subject of the PP and the PP form a CFC -a small clause (see Schein (1982), Stowell (1981, 1991), and Williams (1975, 1983), for different analyses of small clauses). The PP subject need not be within this PP (cf. Giorgi (1984)), as evidenced by the fact that the PP can be moved without its subject, as illustrated in (i):

(i) ɦaddo **la-[kəll walad]$_1$** ʃeefit zeena ʔəmm-[o]$_1$
next-him to-every boy saw.3SF Z. mother-his
'Next to every boy, Zeina saw his mother.'

Since the subject need not form a constituent with the PP, it can bind a pronoun within a doubled element attached to this PP (ii):

(ii) zeena ʃeefit [kəll walad]$_1$ ɦadda **la-ʔəmm-[o]$_1$**
Z. saw.3SF every boy next-her to-mother-his
'Zeina saw every boy next to his mother.'

In (ii), it is possible to assume that the PP contains a non-overt category coindexed with its subject *kəll walad* ('every boy').

[13] The doubled element always appears to the right of the doubling clitic. This is captured by assuming a raising analysis for the complex predicate or the host containing the clitic. One may assume that when no raising of the predicate or the host containing the clitic takes place, the sentence ultimately generated will be a clitic-left-dislocated sentence. In other words, I am surmising that both clitic-left-dislocated structures and clitic-doubling will be generated from the same underlying representation. In Kayne (1994), it is assumed that clitic-doubled constructions underlie clitic-left-dislocated constructions.

[14] In sentences (66a-b), the subject can also be analyzed as a clitic-left-dislocated element. In that case, this clitic-left-dislocated element which is outside the CFC can bind a pronoun contained within this doubled element as in (i):

(i) kəll mʕallme$_1$ χallayto yinʒaɦ **la-təlmiiz-[a]$_1$**
every teacher let.1S-him pass.3SM to-student-her
'Every teacher, I let her student pass.'

It goes without saying that the object QP in (ii) cannot be so analyzed. As such, it cannot bind the doubled element:

(ii) * [χallayto yirsum kəll mʕallme] **la-təlmiiz-[a]$_1$**
let.1S-him draw.3SM every teacher to-student-her
'I let her student draw every teacher.'

The object QP in (ii) is raised with the whole CFC and, as such, does not c-command the doubled element.

[15] In this respect, doubling is similar to the process of clitic-left-dislocation (see Cinque (1990), Iatridou (1991), Demirdache (1991)). It is to be added to the list of mechanisms grammars use to create new arguments, such as benefactive dative constructions, benefactive applicative constructions, etc.... (see Borer and Grodinsky (1986), Baker (1988); see also Vergnaud and Zubizarreta (1992) for relevant discussions concerning the projection of argument structures). It is interesting to note in this connection that doubled elements, like benefactive datives, are marked datives. This may be taken to suggest that dative is a way to mark the occurence of non-conventional subjects. That is, along the lines of Kayne (1984) and Aoun and Li (1989), datives are to be analyzed as secondary-subjects (see footnote 1 for a different suggestion concerning the function of the dative marker *la*).

REFERENCES

Aoun, J.: 1979, 'On Government, Case Marking, and Clitic Placement', ms., MIT, Cambridge, Mass.

Aoun, J.: 1981, *The Formal Nature of Anaphoric Relations*. Doctoral Dissertation, MIT, Cambridge, Mass.

Aoun, J. and N. Hornstein: 1985, 'Quantifier Types', *Linguistic Inquiry 16*, 623-637.

Aoun, J.: 1993, 'The Syntax of Doubled Arguments', *International Journal of Basque Linguistics and Philology 27*, 709-730.

Aoun, J., E. Benmamoun, and D. Sportiche: 1994, 'Agreement, Word Order, and Conjunction in Some Varietes of Arabic', *Linguistic Inquiry 25*, 195-220.

Aoun, J. and Y. A. Li: 1989, 'Constituency and Scope', *Linguistic Inquiry 20*, 124-172.

Baker, M.: 1988, *Incorporation: A Theory of Grammatical Function Changing*. Chicago: University of Chicago Press.

Benmamoun, E.: 1992, *Functional and Inflectional Morphology: Problems of Projection, Representation, and Derivation*. Doctoral Dissertation, USC, Los Angeles, California.

Borer, H.: 1983, *Parametric Syntax*. Dordrecht: Foris.

Borer, H.: 1986, 'The Syntax of Pronominal Clitics', In H. Borer, ed., *Syntax and Semantics 19: The Syntax of Pronominal Clitics*. New York: Academic Press.

Borer, H., and Y. Grodinsky: 1986, 'Syntactic Cliticization and Lexical Cliticization: The Case of Hebrew Dative Clitics', In H. Borer, ed., *Syntax and Semantics 19: The Syntax of Pronominal Clitics*. New York: Academic Press.

Browning, M. A.: 1987, *Null Operator Constructions*. Doctoral Dissertation, MIT, Cambridge, Mass.

Chomsky, N.: 1982., *Some Concepts and Consequences of the Theory of Government and Binding*. Cambridge, Mass.: MIT Press.

Chomsky, N.: 1986, *Knowledge of Language: Its Nature, Origin, and Use*. New York: Praeger.

Chomsky, N.: 1991, 'Some Notes on the Economy of Derivations and Representations', In R. Friedin, ed., *Principles and Parameters in Comparative Grammar*. Cambridge, Mass.: MIT Press.

Chomsky, N.: 1992, 'A Minimalist Program for Linguistic Theory', *MIT Occasional Papers in Linguistics 1*. Cambridge, Mass.

Cinque, G.: 1990, *Types of A'-Dependencies*. Cambridge, Mass.: MIT Press.

Demirdache, H.: 1991, *Resumptive Chains in Restrictive Relatives, Appositives and Dislocation Structures.* Doctoral Dissertation, MIT, Cambridge, Mass.

Dobrovie-Sorin, C.: 1990, 'Clitic-Doubling, Wh-Movement, and Quantification in Romanian', *Linguistic Inquiry 21*, 351-398.

Everett, D.: 1984, 'Piraha: Clitic-Doubling and the Parametrization of Nominal Clitics', ms., MIT and SIL/UNICAMP.

Fiengo, R., and J. Higginbotham: 1981, 'Opacity in NP', *Linguistic Analysis 7.*

Franco, J.: 1991, 'Spanish Object Clitics as Verbal Agreement Morphemes', *MIT Working Papers in Linguistics 14,* 99-114. Cambridge, Mass.

Giorgi, A.: 1984, 'Towards a Theory of Long Distance Anaphors: a GB Approach', *The Linguistic Review 3*, 307-361.

Hurtado, A.: 1984, 'Clitic Chains', Ms., Simon Fraser.

Iatridou, S.: 1991, 'Clitics and Island Effects', ms., MIT, Cambridge, Mass.

Jaeggli, O.: 1982, *Topics in Romance Syntax.* Dordrecht: Foris.

Jaeggli, O.: 1986, 'Three Issues in the Theory of Clitics: Case, Double NPs, and Extraction', In H. Borer, ed., *Syntax and Semantics 19: The Syntax of Pronominal Clitics.* New York: Academic Press.

Kayne, R.: 1975, *French Syntax: The Transformational Cycle.* Cambridge, Mass.: MIT Press.

Kayne, R.: 1984, *Connectedness and Binary Branching.* Dordrecht: Foris.

Kayne, R.: 1994, *The Antisymmetry of Syntax.* Cambridge, Mass.: MIT Press.

Koopman, H.: 1991, 'Restrictions on Spec Positions', ms., UCLA, California.

May, R.: 1985, *Logical Form: Its Structure and Derivation.* Cambridge, Mass.: MIT Press.

Mohammad, M.: 1990, *The Sentence Structure of Arabic.* Doctoral Dissertation, USC, Los Angeles, California.

Rothstein, S.: 1983, *The Syntactic Forms of Predication.* Doctoral Dissertation, MIT, Cambridge, Mass.

Rouveret, A., and J.-R. Vergnaud: 1980, 'Specifying Reference to the Subject', *Linguistic Inquiry 11.*

Schein, B.: 1982, 'Small Clauses and Predication', ms., MIT, Cambridge, Mass.

Schneider-Zioga, P.: 1990, 'Double Object Constructions and Small Clauses', ms. USC, Los Angeles, California.

Schneider-Zioga, P.: 1993, *The Syntax of Clitic Doubling in Modern Greek.* Doctoral Dissertation, USC, Los Angeles, California.

Shlonsky, U.: 1991, 'Quantifiers as Functional Heads: A Study of Quantifier Phrases and Quantifier Float', *Lingua 84*, 159-180.

Sportiche, D.: 1992, 'Clitic Constructions', ms., UCLA, Calilfornia.

Stowell, T.: 1981, *Origins of Phrase Structure.* Doctoral Dissertation, MIT, Cambridge, Mass.

Stowell, T.: 1991, 'Small Clause Restructuring', In R. Freidin., ed., *Principles and Parameters in Comparative Grammar,* Cambridge, Mass.: MIT Press.

Suñer, M.: 1988, 'The Role of Agreement in Clitic Doubled Constructions', *NLT 6*, 391-434.

Suñer, M.: 1992, 'Two Properties of Clitics in Clitic Doubled Constructions', In J. Huang and R. May, eds., *Logical Structure and Linguistic Structure: Cross Linguistic Perspectives*. Reidel.

Vergnaud, J.-R., and M.-L. Zubizarreta.: 1992, 'The Definite Determiner and the Inalienable Constructions in French and English', *Linguistic Inquiry 23*, 595-652.

Williams, E.: 1975, 'Small Clauses in English', In *Syntax and Semantics 4, 249-273,* ed. J.P. Kimball. New York: Academic Press.

Williams, E.: 1980, Predication', *Linguistic Inquiry 11*, 203-238.

Williams, E.: 1983, 'Against Small Clauses', *Linguistic Inquiry 14,* 287-308.

DECONSTRUCTING THE CONSTRUCT*

Hagit Borer
University of Southern California

1. INTRODUCTION

In the recent decade, Construct State nominals (henceforth CSN) in Hebrew and Arabic have been the focus of much discussion and research, resulting in many important insights. Originating with a proposal of Ritter (1987, 1988) many analyses of that construction seek to tie (at least some of) its properties to raising of N to D (cf. Ritter, (1987, 1988, 1991); Fassi Fehri (1988, 1989, 1994); Ouhalla (1990, 1991); Hazout (1990); Muhammad (1988); Duffield (1992); Siloni (in press, 1994); among others). The purpose of this paper is twofold. First, I will review and evaluate the major insights behind the N-to-D hypothesis, showing that while some of them have led to an increased understanding of CSN, the N-to-D hypothesis does not, in fact, achieve the major tasks which its proponents claim it to achieve. I will conclude that while the operation of N-to-D cannot be directly excluded, based on word order, there is no direct evidence for its existence and the word order phenomena which such an account seeks to explain must be accounted for in some other way.

In the second part of the paper I will resurrect the N-to-D analysis, this time not to account for properties of CSN, but rather, as a well-formedness condition on the realization of ±definite features. I will propose that in Semitic languages, ±definiteness is a feature base-generated on the N stem (cf. Borer (1988); Siloni (1994)), while the D head itself is unspecified for such a feature, inheriting it from a moved N. I will further argue that the feature-nature of Definiteness in Semitic allows for the base-generation of a nominal stem without it, and that such a state of affairs leads, by independent principles, to the formation of CSN. The absence of CSN in, e.g., the Germanic languages, correlates with the fact that in these languages, ±definitness is not a feature on the N stem, but rather, a feature of the head D itself.

This paper is organized as follows: In section 2 I discuss some of the properties of CSN, and compare them to those of free nominals (henceforth FN). In section 3 I review the advantages and disadvantages of the preliminary proposals made by Ritter (1987, 1988) and others to account for the properties of CSN. Modifications of these proposals, based on earlier criticism, are reviewed in section 4.

In section 5 I turn to the Embedded VP Hypothesis: the hypothesis advanced in Hazout (1990) and Borer (1991) that process nominals (Complex Event Nominals, in the

K. Johnson and I. Roberts (eds.), Beyond Principles and Parameters. 43 -89.

terminology of Grimshaw (1990)) contain a fully projected VP. This hypothesis in turn will serve as a jumping board for a syntactic comparison between process nominals and non-process nominals, the latter shown to have a clearly different syntactic structure, regardless of whether they are CSN or FN. Given this differing structure, there is no motivation from word order for N raising to D or to any other functional projection.

In section 6 I investigate the word properties of CSN, concluding that it involves syntactic incorporation, and propose a set of conditions on well-formed incorporation, using Grimshaw's (1991) notion of Extended Projections as a stating point.

In section 7 I turn to an analysis of CSN. Assuming that the N head of CSN is not lexically marked as ±definite, and must acquire such marking to become well formed, it is shown that this minimal assumption, coupled with otherwise motivated principles of the grammar, suffices to derive all the properties of CSN, as distinct from those of FN.

2. PROPERTIES OF CSN AND OF FN:

As is quite well-known, CSN have the properties in (1), illustrated in (2)-(7):

(1) **Properties of Construct State Nominals (CSN)**:

A. Strictly right branching, head first, apparent NSO order, (but see sections 5, 6). (2)

B. The head of a CSN cannot be directly modified by a determiner or by an adjective. (3)

C. The ±definite value of the rightmost branch of a CSN is associated with each N in the CSN and the CSN as a whole (henceforth Definiteness Spreading). (4)

D. Adjectival modification of the head or the entire CSN follows the (last) right branch. (5)

E. If more than one N in CSN is modified by an adjective, the configuration is nested. (6)

F. Heads are bound morphemes: they do not bear main stress, and a variety of word-internal phonological rules apply to them. (7)

G. CSN strategy is available for compound formation. (8)

(2) a. [delet [beit [morat [ha-kita]]]]
door house teacher the-class
'the door of the house of the teacher of the class'

b. [harisat [oyvey [ha-miStar]]] 'et ha-'ir
destruction enemies the-regime OM the-city
'the enemies of the regime's destruction of the city'

c. [tipul [Siltonot ['ayarat [ha-mehagrim]]]] ba-ba'aya
treatment authorities town the-immigrants in-the-problem
'the authorities of the immigrants' town treatment of the problem'

(3) a. (*ha-)delet (*ha-)beit (*ha-)morat (ha-)kita
(*the)door (*the)house (*the)teacher (the)class

b. delet (*rexava) beit (*lavan) morat (*vatika) ha-kite (ha-xadaSa)
door (*wide) house (*white) teacher (*senior) the-class (the new)

(4) a. beit more
'a house of a teacher'

b. beit ha-more
'the house of the teacher'
*'a house of the teacher'
*'the house of a teacher'

(5) delet beit morat ha-kite ha-yafa
door-f house-m teacher-f the-class-f the-beautiful-f

a. the beautiful door of the house of the teacher of the class
b. the door of the beautiful teacher of the class
c. the door of the house of the teacher of the beautiful class

(6) a. kis'ot ha-kita ha-xadaSa ha-civ'onim
chairs-m-pl. the-class-f the new-f the-colorful-m-pl.

b. *kis'ot ha-kite ha-civ'onim ha-xadaSa
chairs-m-pi. the-class-f the-colorful-m-pl the new-f

(7) **CSN** **FN**

a. beit more — bayit Sel mora
house teacher — house of teacher
'a teacher's house'

b. morat kita — mora(a) Sel kita
teacher class — teacher of class
'a teacher of a class'

(8) beit xolim
house sick-pl.
'hospital'

beit sefer
house book
'school'

vex din
editor law
'lawyer'

etc.

Many, but not all, of these properties contrast with those exhibited in free nominals (FN), constructed roughly on a par with the English possessive construction, and containing a possessive preposition, *Sel*, mediating between the *possessed* and the *possessor*:

(9) **Properties of FN**:

A. Head first as well, apparent NSO order, but see comment following (IA)). (10)

B. The head must be directly marked for ±definiteness.(10)

C. The ±definite value is strictly associated with the N marked as such (no Definiteness Spreading). (11)

D. Adjectival modification of the head follows the head directly. (12)

(10) a. ha-delet Sel ha-bayit Sel ha-more Sel ha-kita
the door of the house of the-teacher of the-class

b. ha-harisa Sel ha-oyvim Sel ha-miStar 'et ha-'ir
the-destruction of the-enemies of the-regime OM the-city

c. ha-tipul Sel ha-Siltonot Sel ha-'ayara Sel ha-mehagrim
the-treatment of the authorities of the town of the immigrants

ba-ba'aya
in-the-problem

(11) a. ha-bayit Sel ('eyze) mora
the-house of (some) teacher
'the house of (some) teacher'

b. bayit Sel ha-mora
house of the-teacher
'a house of the teacher'

(12) a. ha-delet ha-yafa Sel ha-bayit Sel ha-mora
door-f the-beautiful-f of the-house-m of the-teacher-f

b. ha-delet Sel ha-bayit Sel ha-more ha-yafa
door-f of the-house-m of the-teacher-f the-beautiful-f

3. ADVANTAGES OF THE N-TO-D ANALYSIS.

Ritter (1987,1988) as well as subsequent treatments of CSN are motivated by an attempt to give a unifying account for at least three of the most salient properties of CSN, as listed in (13):

(13) A. The strict right branching nature of the construction
B. The NSO word order
C. The prohibition on direct modification of the head.

In a variety of executions which differ slightly but not fundamentally, it has been proposed that CSNs such as (2b-c) are base-generated in essence as in (14a). In (14a), note, the word order inside the NP is SNO, as is expected. In that structure, N raises to D, accounting for the NSO word order. Further, under the assumption (again, slightly differently executed in different accounts) that the movement of N to D blocks the overt realization of a determiner, the impossibility of direct modification of the head by the definite article follows. The right branching of the construction follows as well, resulting in the S-structure in (14b):

(14) a. [$_{DP}$ [$_{\bar{D}}$ ∅ [$_{NP}$ **ha-oyvim** [$_{\bar{N}}$ **harisat** **'et ha-'ir**]]]]
the-enemies destruction OM the-city

b. [$_{DP}$ [$_{\bar{D}}$ **harisat** [$_{NP}$ **ha-oyvim** [$_{\bar{N}}$ t_N **'et ha-'ir**]]]]
destruction the-enemies OM the-city

The account is prima facie an attractive one, benefitting from the parallelism between the structure of clauses and the structure of nominal projections, and rejecting it should be done only if the properties associated with CSN can be otherwise accounted for. It should be mentioned, however, that two important properties of CSN are not directly accounted for by the structures in (14a-b) and require additional mechanisms: the first is Definiteness Spreading. Addressing this question, Hazout (1990) observes that in English as well, a definite subject of NP causes the entire NP to be interpreted as definite, while an indefinite subject causes the entire NP to be indefinite:

(15) a. John's book (interpreted as a definite book)
b. a student's notebook (interpreted as an indefinite notebook)

Definiteness Spreading in CSN, Hazout reasons, should be explained on a par.

The analogy between the English structure and Hebrew one, however, is not straightforward. Plausibly, the genitive marker *'s* in (15) is the head of D, and *John/student* are in [Spec, DP], thus triggering ±definiteness by Spec-Head agreement, as in the structure in (16):

(16) a. $[_{DP}$**John**$_2$ $[_{\bar{D}}$ **'s** $[_{NP}$ $\boldsymbol{t_2}$ **book**]]]

b. $[_{DP}$ **a student**$_2$ $[_{\bar{D}}$ **'s** $[_{NP}$ $\boldsymbol{t_2}$ **notebook**]]]

In the CSN structure in (14b), on the other hand, the subject of the NP clearly cannot be in [Spec,DP], or the necessary word order configuration would not emerge. Nevertheless, it is possible that an LF operation raises that subject from [Spec,NP] to [Spec,DP], thereby rendering the entire expression ±definite as desired. Such LF raising is explicitly proposed by Fassi Fehri (1994). The resulting LF structure, presumably, is as in (17):

(17) $[_{DP}$ **ha-oyvim**$_2$ $[_{\bar{D}}$ **harisat** $[_{NP}$ $\boldsymbol{t_2}$ $\boldsymbol{t_N}$ **'et ha-'ir**]]]
the-enemies destruction OM the-city

A serious drawback to the LF raising account, however, is the ungrammaticality of (18a-b) in English:

(18) a. *$[_{DP}$ sunflower$_2$ $[_{\bar{D}}$'s $[_{NP}$ picture t_2]]]

b. *$[_{DP}$D the subject$_2$ $[_{\bar{D}}$'s $[_{NP}$ learning t_2]]]

For reasons that remain rather poorly understood, a *depicted* in a 'picture' NP, or a non-affected object cannot find themselves in [Spec,DP]. As a result, no Definiteness Spreading of any sort is ever attested in these cases, and (19)-(20) are fully grammatical. Note, crucially, that in the absence of Definiteness Spreading in (19)-(20), an LF movement raising *sunflowers* or *the subject* (in English) to [Spec, DP] is unlikely (note that as *learning* is a process nominal, an indefinite article is barred independently, as are all count determiners. *Some* is substituted):

(19) a. [$_{DP}$ the [$_{NP}$ picture of the sunflowers]]
b. [$_{DP}$ a [picture of a sunflower]]
c. [$_{DP}$ the [picture of a sunflower]]
d. [$_{DP}$ a [picture of the sunflower]]

(20) a. [$_{DP}$ the [$_{NP}$ learning of the subject]]
b. [$_{DP}$ some [learning of a (difficult) subject]]
c. [$_{DP}$ the [learning of a subject]]
d. [$_{DP}$ some [learning of that subject]]

In contrast with the situation in (19)-(20). Definiteness Spreading in Hebrew does apply in the analogous situations:

(21) a. tmunat ha-xamaniyot
'the picture of the sunflowers'

b. tmunat xamaniyot
'a picture of sunflowers'

(22) a. limud ha-nose
'the learning of the subject'

b. limud nose
'learning of a subject'

Thus in order to maintain the LF movement analysis for Definiteness Spreading not only would we have to assume LF movement in Hebrew where it is not attested overtly, on a par with overt movement in English, but also LF movement where both overt and LF movement are blocked in English. Clearly, then, trying to reduce the Semitic effect to the English paradigm is problematic.

A second issue which remains unaddressed directly by (14a-b) concerns the placement of modifying adjectives. In regular free nominals, adjectival modifiers appear immediately after the N and preceding its complements, as (23) exemplifies:

(23) ha-more ha-ce'ira le-fizika
the-teacher the-young for physics

The placement of adjectives in regular FN is already problematic. Adjoining APs to $\overline{\mathrm{N}}$ or higher is necessary in English as well as in Hebrew, to allow recursive modification with its well established scopal properties. In (24a) *second* has scope over *successful [candidate for the job]* as is the case in (25b):[1]

(24) a. the [second [successful [candidate for the job]]]
b. the [successful [second [candidate for the job]]]

(25) a. ha-muamedet ha-Sniya ha-muclaxat la-micra
the-candidate the-second the-successful for the job
'the successful second candidate'

b. ha-muamedet ha-muclaxat ha Sniya la-misra
the-candidate the-successful the-second for the job
'the second successful candidate'

But if APs are adjoined at the $\overline{N}$ level (or higher) and no further movement is attested, such adjunction would give rise to (26a), for left adjunction, and to (26b), for right adjunction, both ungrammatical:

(26) a. $[_{DP}$ **ha** $[_{NP}$ $[_{\overline{N}}$ **ha-xadaSa** $[_{\overline{N}}$ **mora le-fisika**]]]]
the the-new teacher to-physics

b. $[_{DP}$ **ha** $[_{NP}$ $[_{\overline{N}}$ $[_{\overline{N}}$ **mora le-fisika**] **ha-xadaSa**]]]
the teacher to-physics the-new

Interestingly, an $\overline{N}$ (or higher) left-adjunction site for the adjective can be maintained, if we assume that in a structure such as (26b), the head N. *mora* 'teacher' raises to D. However, such N-to-D raising has been proposed for CSN and not for FN, in order to account for contrasts such as (5) on the one hand, and (I 2a) on the other hand.

N-to-D movement in CSN does go some way toward explaining the distribution of APs in CSN. Thus consider (27):

(27) a. $[_{DP}$ $[_{NP}$ kita $[_{\overline{N}}$ xadaS $[_{\overline{N}}$ more]]]]
class-f new-m teacher-m

b. $[_{DP}$ more$_1$ $[_{NP}$ kita $[_{\overline{N}}$ xadaS $[_{\overline{N}}$ t_1]]]]
teacher class new

'a new teacher of a class'

But when a slightly more complicated CSN structure is considered, in which both head and complement are modified, the explanatory power of N-to-D movement disappears: the word order of adjectives (cf. 28c) cannot be derived. Rather, the ungrammatical (28b) is derived:

(28) a. $[_{DP1}$ $[_{NP1}[_{DP2}$ $[_{NP2}$ $[_{N'2}$ mitkademet kita $[_{\bar{N}1}$ xadaS $[_{\bar{N}1}$ more]]]]
advanced-f class-f new-m teacher-m

b. *$[_{DP1}$ more$_1$ $[_{NP1}[_{DP2}[_{NP2}[_{\bar{N}2}$ mitkademet kita $[_{\bar{N}1}$ xadaS $[_{\bar{N}1}$ t_1]]]]
teacher-m advanced-f class-f new-m

c. more kita mitkademet xadaS
teacher class advanced new

'a new teacher of an advanced class'

4. ADDITIONAL PROBLEMS, SOME MODIFICATIONS

A more serious problem specifically for the structures in (14a-b) results from the fact that FN utilizing the genitive preposition *Sel* 'of' exhibit an identical NSO word order as already exhibited by (10b-c).

Clearly, the word order in (lOb-c), identical to that found in CSN, is independent of the other -properties associated with CSN: the positioning of modifiers and Definiteness Spreading. FN do not exhibit Definiteness Spreading, and adjectives are placed directly after the head they modify, preceding PP complements as well as *Sel*-phrases, as (29a-d) exemplify:

(29) a. ha-harisa-f ha-'axzarit-f Sel ha-'oyev-m ha-xamuS-m 'et ha-'ir
the-destruction the-cruel of the-enemy the-armed OM the-city

b. ha-sefer ha-xadaS 'al ha-Stixim ha-parsim
the-book the-new about the rugs the Persian

c. ha-sefer 'al ha-Stixim Sel ha-sifriya
the-book about the-rugs of the-library

Clearly, then, a single N-to-D movement deriving properties (13A,B,C) is untenable. Ritter (1991), approaching this problem directly, thus suggests that two types of head movement and two distinct functional projections are associated with nominals in Hebrew. The first head movement, attested in all nominals, is N-to-Num (=number), head of NumP, resulting in NSO word order for CSN and FN alike. A second movement, unique to CSN, is Num-to-D, accompanied by the movement of the subject of NP from [Spec,NP] to [Spec,NumP]. This latter movement is intended to derive Definiteness Spreading and the placement of adjectives. A somewhat similar solution is proposed by Siloni (1994). Siloni proposes that all N heads, in both CSN and FN, move to D to check their ±definite base-generated features, adopting a proposal in Borer (1988), that the definite article in Hebrew is a base-generated feature on the head N. Although heads of CSN are not overtly marked as ±definite, Siloni assumes that they are nevertheless inherently marked as ±definite, and that this marking is realized as the special bound morphology on the heads of CSN. As a result, such nouns must move to D to check their features. As heads of CSN are also inherently marked with agreement features, these features are checked in an intermediate functional head, AgrGen, whose Specifier, in turn, assigns genitive Case to the subject of the NP moved there. The derivations proposed by Ritter (1991) and Siloni (1994) are given in (30)-(31) respectively:

(30) FN: [$_{DP1}$ **ha-**[$_{NumP1}$**harisa** [$_{NP1}$ [$_{NP1}$[$_{DP2}$(Sel) **ha-**[$_{NumP2}$ **'oyev** [$_{NP2}$
the destruction-f ⇑ of the enemy-m ⇑
AP AP
ha-'axzarit **ha-xamuS**
the-cruel-f the-armed-m

[$_{NP2}$ t_{n2}]]]]$_{tn1}$ **'et ha- 'ir]]]]**
OM the-city

CSN: [$_{DP}$ **harisat** [$_{NumP}$ [$_{DP}$ **ha-**[$_{NumP}$ **`oyev** [$_{NP}$ [$_{NP}$ $t_{n2}$]]]]][$_{NP}$ [$_{NP}$ $t_2 t_n$ **`et ha-`ir]]]**
destruction-f the enemy ⇑ ⇑ OM the-city
AP AP
ha-xamuS **ha-'axzarit**
the-armed-m the-cruel-f
(Ritter, 1991)

(31) FN: [$_{DP}$**ha-harisa** [$_{NP}$ [$_{NP}$[$_{DP}$ **(Sel) ha-'oyev** [$_{NP}$ [$_{NP}$ t_{n2}]]] t_n **'et ha-'ir]]**
the destruction-f ⇑ of the enemy-m ⇑ OM the-city
AP AP
ha-'axzarit **ha-xamuS**
the-cruel-f the-armed-m

CSN: [$_{DP}$**harisat** [$_{AGRgen}$ [$_{DP}$ **ha-'oyev** [$_{NP}$ [$_{NP}$ t_{n2}]]] t_n [$_{NP}$ [$_{NP}$ t_2 t_n **'et ha-'ir]]]**
destruction the-enemy ⇑ ⇑ OM the-city
AP AP
ha-xamuS **ha-'axzarit**
the-armed-m the-cruel-f
(Siloni, 1994)

As an advantage of both analyses, note that they account for the nested distribution of AP modifiers. Suppose, as has been recently proposed by Valois (1991), AP modifiers are adjoined to NP. As the structures in (30)-(31) both involve movement of N outside the NP projection, we expect AP to always follow N. As both analyses also involve movement of the subject of N outside the NP in CSN, but not in FN, the placement of the AP after this subject in CSN follows as well. The placement of APs in these analyses is marked with arrows in (30)-(31).

5. WORD ORDER IN NOMINALS: PROCESS NOMINALS VS. NON-PROCESS NOMINALS

Hazout (1988, 1990, 1995) as well as Borer (1991, forthcoming) argue that word order in **process nominals** (Grimshaw's 1990 **Complex Event Nominals**), as exemplified specifically in (2b,c), but not in underived nominals or non-process derived nominals, is derived not from the raising of N-to-D, but rather, from the existence of an embedded VP, and the raising of V to an N head. Abstracting away from some differences between Hazout's analysis and Borer's analysis at this point, the D-structure for process nominals in FN configurations such as (32a-b) would be as in (33a). The derived structure is given in (33b):

(32) a. ha-harisa Sel ha-'oyev 'et ha-'ir
the-destruction of the-enemy OM the city

b. ha-tipul Sel ha-Siltonot ba-ba'aya
the-treatment of the authorities in the problem

(33) a. $[_{DP}$ $[_{NP}$ N $[_{VP}$ Subj V OM Obj/PP]]]

b. $[_{DP}$ $[_{NP}$ V+N $[_{VP}$ Subj t_V OM Obj/PP]]]
harisa (Sel) ha-oye et ha-'ir
destruction of the enemy OM the-city

Although the analyses proposed for CSN in Hazout (1990) and Borer (1991) differ, both share the assumption that the structure in (33) is the structure which serves as an input to the CSN in (34):

(34) a. harisat ha-oyev 'et ha-'ir
destruction the-enemy OM the city

b. tipul ha-Siltonot ba-ba'aya
treatment the-authorities in-the-problem

The most salient difference between the Embedded VP Hypothesis and the accounts presented above is that for proponents of the former, the derivation of word order in process nominals is entirely independent from the existence of functional projections or from any properties associated with CSN. While the Embedded VP Hypothesis borrows from Ritter's original intuition the assumption that the word order is derived by the extraction of a head from the clause it heads, the movement here is of an embedded V head, incorporating onto a dominating N head, thereby resulting in NSO word order. The formation of CSN nominals, in turn, is an independent operation on the structure in (33), bringing together the output of V+N with the first constituent of the embedded VP, in the case of (33), the embedded subject. Borer (1991) further makes a principled distinction between process nominals, in which arguments are present which are part of the argument structure of the embedded V, and non-process nominals, whether derived or underived, in which a VP is not present, and hence, there is no argument structure as such (although, of course, *possessor* or PP modifiers are possible).

Putting aside here some criticism of the Embedded VP Hypothesis[2] an important question can now be raised. If the NSO word order in process nominals derives from V-to-N raising, what accounts for the N-initial position in NPs which are not process nominals? Specifically, what accounts for the word order in (2a), or in NPs such as those in (35):

(35) a. ha-sefer Sel ha-sifriya 'al ha-Stixim
the-book of the-library about the-rugs

b. ha-tmuna Sel van gogh Sel ha-xamaniyot
the-picture of Van Gogh of the sunflowers

It turns out, however, that there are some crucial differences between the non-process nominals NPs in (35), and process nominals. While in cases such as (35) the NP must always be head initial, Shlonsky (1988) observes that the order of the post-head elements is free. Thus alongside (35a-b) we have (36a-c), with identical interpretation:

(36) a. ha-sefer' al ha-Stixim Sel ha-sifriya
the-book about the-rugs of the library

b. ha-tmuna Sel ha-xamaniyot Sel van gogh
the-picture of the sunflowers of Van Gogh

c. ha-tmuna Sel ha-xamaniyot Sel ha-muzeon
the picture of the sunflowers of the museum[3]

In contrast, no such freedom of word order is available for process nominals. These come in two varieties, each with a strict word order. An active variety, as in (2b), repeated in (37), with an alternation in word order leading to ungrammaticality, as (38a-b) illustrate, and a passive variety, in which the N head is followed by the object and the subject is expressed as a *by*-phrase, as in (39) (a derivation excluded with a PP complement). Here, too, word order is rigid, as the ungrammaticality of (40a-b) clearly illustrates:

(37) a. ha-harisa **Sel** ha-oyev **'et** ha-'ir
the-destruction of the enemy OM the-city

b. ha-tipul **Sel** ha-Siltonot **ba**-ba'aya
the-treatment of the authorities in the problem

(38) a. *ha-harisa **'et** ha-'ir **Sel** ha-oyev
(*ha-harisa **Sel** ha-'ir **'et** ha-oyev)
(*ha-harisa **Sel** ha-'ir **Sel** ha-oyev)
b. *ha-tipul **ba**-ba'aya **Sel** ha-Siltonot
(*ha-tipul **Sel** ha-ba'aya **Sel** ha-Siltonot)

(39) ha-harisa **Sel** ha-'ir **'al yedey** ha-'oyev
the-destruction of the city by the enemy

(40) *ha-harisa **'al yedey** ha-'oyev **Sel** ha-'ir.
*ha-harisa **Sel** ha-oyev **'al yedey** ha-'ir

The rigidity of word order exhibited in (37)-(38) follows directly from the embedded VP analysis. Suppose the order of the embedded subject and object is determined by the constituent structure within the VP. In (37), the subject is in [Spec,VP], while the direct object or the PP complement are sisters of V. The ungrammaticality of (38) follows directly from the nonavailability of a position into which an NP or a PP complement can be fronted. As for (39), I argue in Borer (1991, forthcoming) that its derivation involves passive inside the embedded VP, with the NP object moved to [Spec,VP], and the embedded subject of the verb expressed as a *by*-phrase.[4] The proposed structure of passive process nominals is as in (41):

(41) a. $[_{DP}$ $[_{NP}$ **N** $[_{VP}$ **V NP**] *by* **NP**]]
b. $[_{DP}$ $[_{NP}$ **V+N** $[_{VP}$ $\mathbf{NP}_1$ $\boldsymbol{t_V t_1}$] *by* **NP**]]

harisa (Sel) ha-'ir 'al yedey ha-'oyev
destruction of the city by the-enemy

If this analysis is correct, we do not expect freedom of word order in (39) any more than we expect it in the comparable sentential passive case. The NP object having moved to [Spec, VP], it must precede the embedded subject of the VP expressed as a *by*-phrase, which is, presumably, right adjoined to the VP.

Crucially, and independently of the Embedded VP Hypothesis, any account for the ungrammaticality of (38), (40) must be based on the assumption that the post-posing of the DP constituent immediately following the N head in process nominals is not possible. This is regardless of whether such a constituent is in [Spec,VP], [Spec,NP], as Ritter (1987, 1988), [Spec,NumP], as in Ritter (1991), or [Spec,Agr_{gen}], as in Siloni (1994). Such post-posing, if allowed, would rule in (38), (40), contrary to fact: it would allow either the subject constituent in (38) or the object constituent in (40) to be post-posed, predicting, erroneously, the grammaticality of NOS orders in active process nominals, and NSO orders in passive process nominals.

But consider now the freedom of word order exhibited in (35)-(36), non-process nominals, when compared with the absence of such freedom in process nominals. For proponents of the Embedded VP Hypothesis a principled structural distinction between process nominals and non-derived nominals follows directly from the presence of a VP in the former vs. its absence in the latter. For opponents of the Embedded VP Hypothesis, on the other hand, in process nominals and non-derived nominals alike, all arguments are associated with an N head: the subject, whether *agent* or *possessor*, is in [Spec,NP] and the complement, whether a *patient*, a *depicted*, or a PP modifier, is a sister of N. Within such analyses, it is rather hard to make a principled distinction between the freedom of word order in non-derived nominals and its rigidity in process nominals. Why would it be possible to place a PP complement such as *'al ha-Stixim* 'about the rugs' preceding the subject, *ha-sifriya* 'the library', in [Spec,NP] as in (36a), but such freedom of word order be blocked in the process nominals (38b), where placing a PP complement *ba-ba'aya* 'in the problem' preceding the subject *ha-Siltonot* 'the authorities' leads to ungrammaticality? Under the assumption that process nominals and non-process nominals have an identical structure the contrast between (36a) and (38b) becomes a complete mystery.

Making matters worse, consider the following facts, observed (but not explained) in Siloni (1994) (see also Shlonsky 1988). Siloni notes that while non-process nominals display the word order freedom illustrated in (35)-(36), this is only the case in FN. In CSN, on the other hand, very strict word order restrictions emerge. Thus, of the combinations in (35)-(36) (and recall that PP complements cannot participate in CSN), only the ones in (42) are licit. The CSNs in (43) are ungrammatical:

(42) a. tmunat ha-xamaniyot Sel van gogh
picture the sunflowers of Van Gogh

b. tmunat ha-xamaniyot Sel ha-muzeon
picture the sunflowers of the museum

(43) a. *tmunat van gogh Sel ha-xamaniyot
the-picture of Van Gogh of the sunflowers

b. *sifrey ha-silriya 'al ha-Stixim
books the-library about the-rugs

c. *tmunat ha-muzeon Sel ha-xamaniyot
picture the-museum of the sunflowers

The generalization is that in the presence of complements in non-process nominals, CSN formation with *possessors* (or understood *agents*) is blocked.[5] If the complement is itself a DP, N+complement CSN formation is possible. If it is a PP, CSN formation is altogether blocked. Now this contrasts sharply with the situation in process nominals. Here, in the presence of both subject and object, and in the active configuration, the only allowed CSN is head+Subj., and never head+Obj. The presence of a subject alone is altogether ungrammatical in both CSN and FN.[6] CSNs formed with the understood complement of process nominals are only possible if the external argument is expressed as a *by* phrase (or is missing altogether), and in line with a passive derivation, the "internal argument" becomes a derived subject:

(44) a. *harisat ha-'ir Sel ha-oyev
destruction the-city of the enemy

b. *harisat ha-'oyev Sel ha-'ir
enemy of the city

(45) a. harisat ha-'oyev 'et ha-ir
destruction the enemy OM the-city

b. harisat ha-'ir ('al yedey ha-'oyev)
destruction the city (by the enemy)

Again, within an approach which assigns to both process nominals and non-process nominals the structure, roughly, in (46), it is difficult to see how these facts can be accounted for. A rather ad hoc restriction would have to block the movement of the subject to [Spec,FP] in (46b), but allow the movement of the object to that position, while the reverse must hold for (46a):

(46) a. $[_{FP1}$ **N** ...$[_{FP2}$... $[_{NP}$ **Subj** t_N **Obj**]]]
harisa 'oyev 'ir
destruction enemy city

b. $[_{FP1}$ **N**... $[_{FP2}$... $[_{NP}$ **Subj** t_N **Obj**]]]
tmuna van gogh xamaniyot
picture van gogh sunflowers

A solution appears available, however, for proponents and opponents of the Embedded VP Hypothesis alike. Suppose *possessors* and *agents* of non-process nominals (but not subjects of process nominals) are not, actually, in [Spec,NP], but rather, they are right-adjoined to NP. Suppose further that movement from such an adjunction site to [Spec,FP_2] is blocked. It thus follows that only an object may move to that position, resulting in the pattern in (47):

(47) a. $[_{FP1}$ **N**...$[_{FP2}$... $[_{NP}$ **Subj** t_N **Obj**] **process nominals**

b. $[_{FP1}$ **N**...$[_{FP2}$... $[[_{NP}$ t_N **Obj**] **Subj**] **non-process nominals**

An auxiliary assumption, to account for the (un)grammaticality in (35)-(41) would be that objects of non-process nominals can be post-posed, hence accounting for the NSO order.

Some independent evidence for a hierarchical structure inside non-process nominals, and for the presence of object post-posing in these structures comes from the following paradigm, adapted from Shlonsky (1988):

(48) a. ha-tmuna Sel kol 'em$_2$ Sel bn-a$_2$
the-picture of every mother of her son

b. ha-tmuna Sel kol ben$_2$ Sel 'im-o$_2$
the-picture of every son of his mother

c. ha-tmuna Sel im-o$_2$ Sel kol ben$_2$
the-picture of his-mother of every son

d. ha-tmuna Sel kol cayar$_2$ Sel kalb-o$_2$
the picture of every painter of his dog

e. *ha-tmuna Sel kol kelev$_2$ Sel ha-cayar Sel-o$_2$
the-picture of every dog of the-painter his

f. ha-tmuna Sel kalb-o$_2$ Sel kol cayar$_2$
the-picture of his dog of every painter

g. *ha-tmuna Sel ha-cayar$_2$ Sel-o$_2$ Sel kol kelev$_2$
the picture of the painter his of every dog
Following Shlonsky (1988)

Shlonsky observes that although the order of the *possessor* and the complement is interchangeable, the interpretation is restricted, allowing only the *possessor* to serve as an antecedent for the complement, but not the other way around. This paradigm thus suggests that the *possessor* asymmetrically c-commands the complement in both NSO and NOS word orders. Now let us combine this conclusion with the conclusion just reached, that the *possessor* is not in [Spec,NP], or the formation of N+Subj. CSN would be erroneously predicted. Under the assumption that the complement is generated as a sister of N. it follows that the *possessor* must be adjoined to the right, and an NSO order derived by rightward adjunction of the object, which does not seem to impact scope.

Let us then conclude that there is strong evidence for generating process nominals with the subject preceding and c-commanding the object, and equally strong evidence for generating nonprocess nominals with the complement preceding the subject linearly (although it is c-commanded by it). This conclusion, however, has a clear consequence: the neutral, non-derived word order in non-process NPs in Semitic involves an N-initial projection. If, however, this is the case, then it follows that movement of N to a functional projection in order to derive the N-initial nature of nominals in Semitic languages is redundant. Regardless of whether such movement is or is not otherwise motivated, it has no effect on word order. The word order in non-process nominals already is N-initial, making the word order motivation for that movement questionable.[7]

Returning to the word order motivation for N-to-D movement, note that such movement would still be necessary to derive NSO configurations for process nominals if the Embedded VP Hypothesis is rejected. Given, however, the non-uniform nature of NP structure to begin with, and the structural differences between process nominals and non-process nominals, one is led to question the assumption that they have anything in common. While the Embedded VP Hypothesis appears, at first sight, to be complex, postulating two completely distinct structures associated with what are, after all, NPs which have much in common, given that these distinctions do, in fact, exist, the postulation of distinct structures, one involving a VP and the other not, seems appealing.

Finally, consider again the behavior of DP complements in process nominals vs. their behavior in non-process nominals. Recall that NSO orders in non-process nominals are assumed to be derived by postposing the complement and adjoining it to the right of the already adjoined *possessor/agent*, a postposing which does not affect c-command. This postposing, however, is blocked in process nominals, correctly excluding (42b). In fact, in process nominals an object may never follow an adjunct:

(49) a. ha-'axila Sel Dan 'et ha-tapu'ax be mehirut / davka'axSav
the-eating of Dan OM the-apple quickly / exactly now

b. ?*ha-'axila Sel Dan be-mehirut / davka'axSav 'et ha-tapu'ax
the-eating of Dan quickly / exactly now OM the-apple

The marginality of (49b) contrasts with the grammaticality of similar configurations in sentential VPs:

(50) a. Dan 'oxel 'et ha-taputax be-mehirut / davka 'axSav
Dan eats OM the-apple quickly / exactly now

b. Dan 'oxel be-mehiru / davka 'axSav 'et ha-tapu'ax
Dan eats quickly / exactly now OM the-apple

But the grammaticality of (50b), when contrasted with (49b), results from the subject of the VP raising to the Specifier of some higher functional projection, while the verb raises to a functional head, as is conventionally assumed (see Emonds (1978), Pollock (1989)). Consider now the derivation of the process nominals in (49b). Although it is assumed within the Embedded VP Hypothesis that the verb has raised to N, the subject of the embedded VP remains in [Spec,VP]. Under the assumption that adverbs are adjoined to maximal projections, the occurrence of an adverb between the VP-subject and the VP-complement would either require $\overline{V}$ adjunction, presumably non-existent, or object post-posing. As object post-posing would have rescued the derivation, rendering it grammatical, the ungrammaticality of (49b) strongly suggests that object post-posing from within the VP is blocked.

Interestingly, (49b) improves greatly if rather than EN, CSN appears, as in (51):

(51) 'axilat Dan be-mehirut 'et ha-tapu'ax
davka 'axSav

eating Dan quickly OM the-apple
exactly now

I will return to the reason for this contrast in section 7.2 below, when I elaborate further on the properties of CSN.

6. CSN AND INCORPORATION

6.1 Hazout, 1990

It follows from the discussion in the previous sections that N-to-D movement, or for that matter, N-to-F movement, if indeed present, cannot be seen as deriving NSO word orders. Such word orders are present, without object postposing, only in process nominals, where they are plausibly derived by V-to-N incorporation. In non-process nominals, the D-structure word order is NOS, and N-to-F raising would not have any impact on it. NSO word orders for such nominals are derived by postposing the complement, which is blocked in VPs.

Once word-order effects are set aside, note that the distinct structures given to process nominals and non-process nominals do not provide an account for the specific properties of CSN when compared with FN. These properties, word order aside, are repeated in (52):

(52i) **Properties of CSN**:
A. Definiteness Spreading
B. No direct modification for the head
C. Nested order of adjectives
D. Word properties

Thus Hazout (1990), although assuming the Embedded VP Hypothesis, potentially resulting in NSO word order in process nominals, nevertheless assumes that N moves to D in CSN. Specifically, Hazout (1990) assumes that the D may dominate either the definite article *ha-* in Hebrew (or *l-* in Arabic) or alternatively, dominate a marker which is inherently marked as ±definite, and which is an assignor of genitive Case -- under government -- to the Specifier of its NP complement. Crucial to his analysis is the assumption that DP complements to CSN heads are interpreted as *possessors*, and that as such, they must occupy the [Spec,NP] position, dedicated to *possessors*. The head of the NP which is being 'possessed' is generated to the right, in a typical $\overline{X}$-theoretic head position, and comes to precede the *possessor* in [Spec,NP] as a result of N-to-D raising. Following such raising, the trace of N becomes associated both with the definiteness value of poss in D and with its genitive assignment features, resulting in the DP in [Spec,NP] being genitive. Hazout's D-structure of (non-process) nominals is given in (53) (note that Hazout postulates APs adjoined to $\overline{N}$ on the right):

(53)

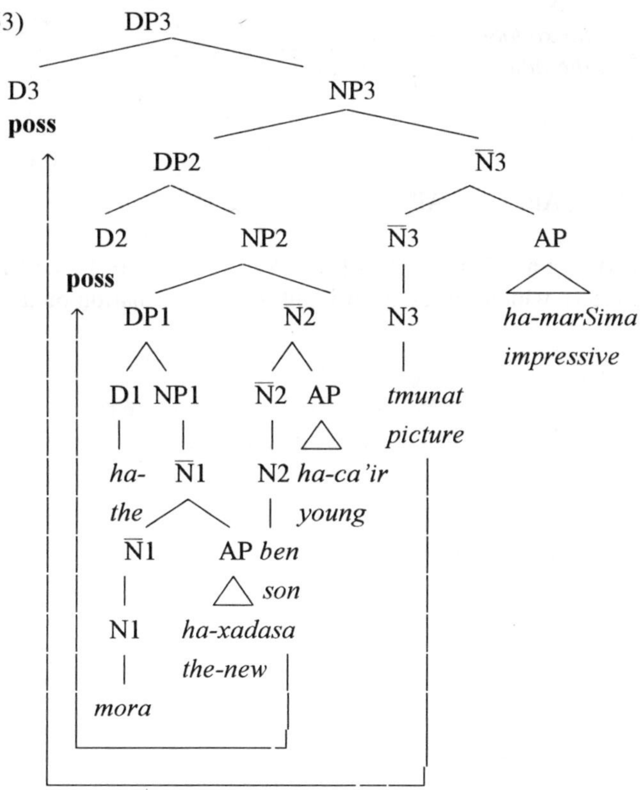

Following the raising of N to D, the resulting structure is as in (54):

(54)

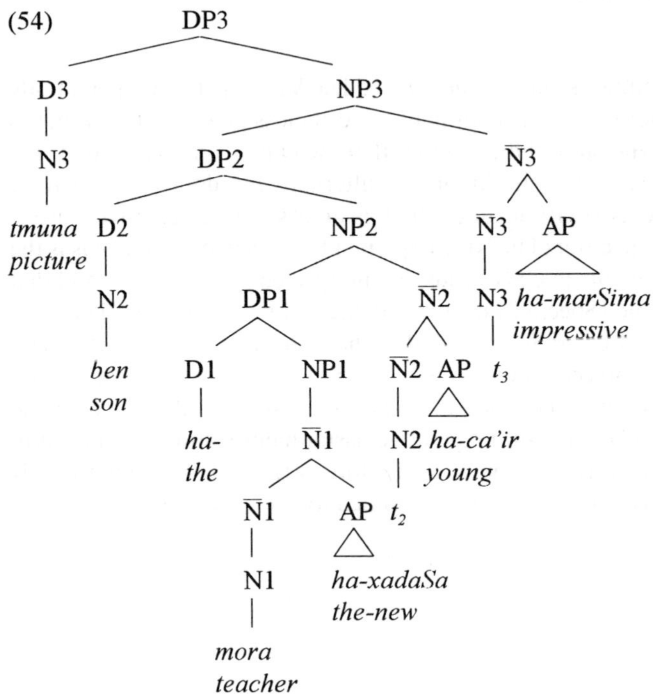

N_3 N_2 N_1 AP_1 AP_2 AP_3

In process nominals, embedding a VP, the DP in [Spec,VP] is raised to [Spec,NP], creating a structure akin to (53), which serves as an input to CSN formation on a par with (54):

(55)

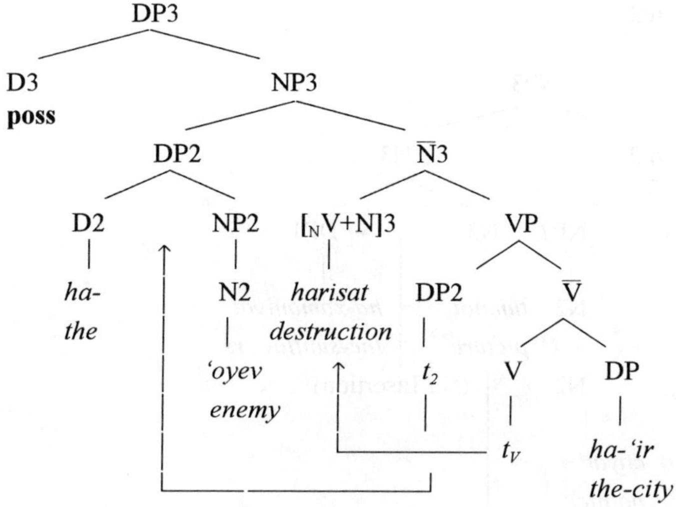

There are a number of clear advantages to Hazout's proposal. First, note that explanation for the complementarity of determiners and CSN constructions also accounts for Definiteness Spreading. A second advantage concerns the placement of adjectives in CSN. The derived word order in (54) is appropriately nested: N3, N2, N1, AP1, AP2, AP3.[8]

Finally, the placement of the understood *possessor* in [Spec,NP] in structures such as (54) captures the subject-like properties of that DP.

Unfortunately, some of the advantages of Hazout's system are mirrored by disadvantages. Difficulties associated with Hazout's idea that Definiteness Spreading in Hebrew is akin to the English definiteness agreement between *possessors* and *possessed* were already discussed (see examples (15)-(22) and related discussion).

A second problem concerns the observations made in section 4. Recall that in the presence of both *possessor/agent* and a complement non-process nominals, CSN formation is only possible with the complement The structures in (53)-(54) make exactly the opposite prediction, as for Hazout, *possessor/agents* are crucially in [Spec, NP], giving them 'priority' in CSN formation (recall that this is precisely the case in process nominals, but not in non-process nominals). Thus the structure of a non-process nominal with both *possessor/agent* and *depicted* complement in Hazout's system would be (presumably) as in (56):

(56)

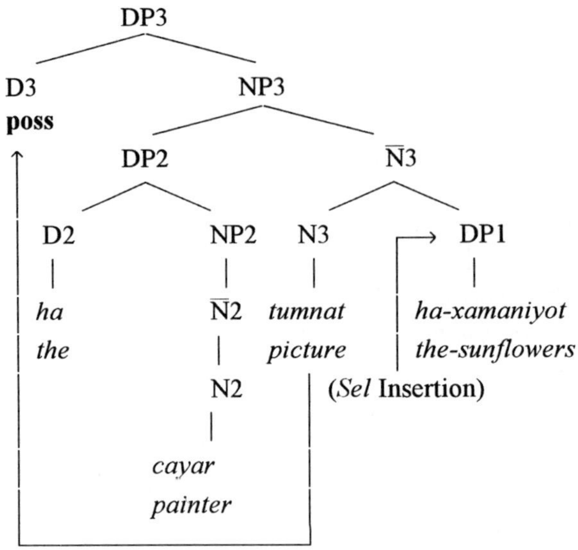

The structure in (56) predicts the grammaticality of *tmunat ha-cayar Sel ha-xamaniyot*, 'picture the painter of the sunflowers' and the ungrammaticality of *tmunat ha-xamaniyot Sel hacayar*, 'picture the sunflowers of the painter', contrary to fact.

Finally, while the structures in (53)-(54) are successful in deriving word order of nouns in CSN, they derive the wrong word order, of both heads and AP modifiers, for FN. (57) is a slight modification of (53) accommodating the properties of FN:

(57)

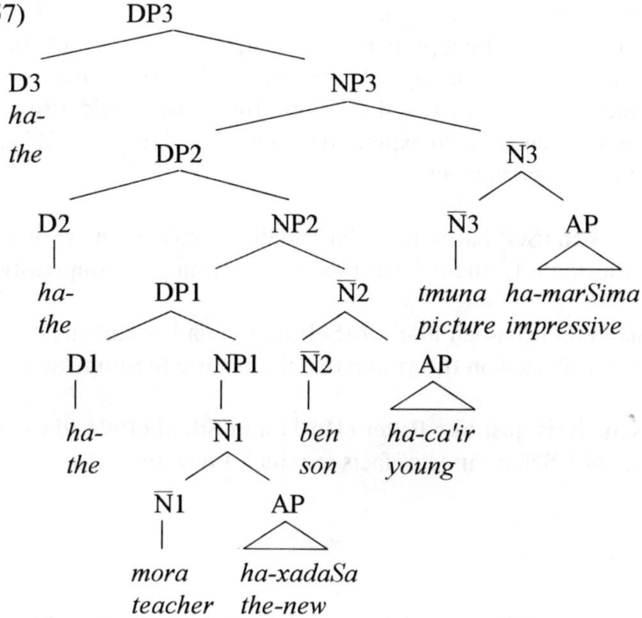

det det det N1 AP1 N2 AP2 N3 AP3

Without any further modification, the structure in (57) would give rise to the ungrammatical sequence in (58):

(58) *ha ha ha more ha-xadaSa ben ha-ca'ir tmuna ha-marSima
the the the teacher the-new son the-young picture the-impressive

There are (at least) three things wrong with (58). First, the stranded definite articles *ha-*, which is a bound morpheme, and its triple concatenation. Second, the order of Ns is exactly opposite to the grammatically-required order (recall that in this structure, it is *picture* which is head, and the *teacher's son* is its possessor), and finally, the order of the APs is exactly opposite of that grammatically required.

Recall that in FN, APs follow immediately the head which they modify. The correct version of (57) is as in (59):

(59) ha-tmuna ha-marSima (Sel) ha-ben ha-ca'ir (Sel) ha-more ha-xadaSa
the-picture the-impressive (of) the-son the-young (of) the-teacher the-new

Consider some ways of modifying the structure in (57) so as to avoid the result in (58). First, the bound morpheme *ha-* and the possibilities of attaching it to its host. It could be assumed that all Ns raise to Ds and adjoin to *ha-*, just as in CSN. There are, however, two problems for Hazout with such a move. First, the resulting word order, as in (60a), would still have the wrong order of APs. Second, if N-to-D occurs in the presence of *ha-*, it is not clear why CSNs display the complementarity between bare N heads and definite articles which is one of their salient properties. Another possibility would be to assume that a post-syntactic morpho-phonological rule attaches *ha-* to its host. Apart from generating an incorrect word order as well (cf. 60b), this move would also cast serious doubt on any account which seeks to explain the complementarity of CSN and definite articles by syntactic head movement.

(60) a. *ha-tmuna (Sel) ha-ben (Sel) ha-more ha-xadaSa ha-ca'ir ha-marSima
the-picture (of) the-son (of) the-teacher the-new the-young the-impressive

b. *ha-more ha-xadaSa (Sel) ha-ben ha-ca'ir (Sel) ha-tmuna ha-marSima
the-teacher the-new (of) the-son the-young (of) the-picture the-impressive

Lastly, note that Hazout's analysis, just like Ritter (1991) and Siloni (1994) does not capture the word properties of CSN, a range of facts to which I now turn.

6.2 CSN as a Word

While the phonological word properties associated with CSN are beyond dispute (see, e.g., Prince (1973)), one could argue that these are the result of post-syntactic processes, a liaison of some sort, which is not syntactically relevant. If this, indeed, is the case, syntactic accounts of CSN which do not consider that aspect are justified in doing so. Consider, however, the examples in (61)-(62), discussed in Borer (1984), as well as the contrast in (63):

(61) a. roS Sel SloSa anaSim nir'a mi-ba'ad la-xalon
head of three people was-seen through the window

b. *roS SloSa anaSim nir'a mi-ba'ad la-xalon
head-three people was-seen through the window

(62) a. tmuna Sel SloSa anaSim hayta munaxat 'al ha-Sulxan
picture (1/3) of three people was lying on the table

b. tmunat SloSa anaSim hayta munaxat 'al ha-Sulxan
picture (1/*3) three people was Iying on the table

(63) a. lo ra'iti tmuna Sel 'iS
not-saw-I picture of anyone

b. *lo ra'iti tmunat'iS
not saw-I picture anyone

Note that, e.g., (61b) cannot be ruled out as morpho-phonologically infelicitous, as *tmunat SloSa 'anaSim*, 'picture three people', as in (62b), is well formed, showing that

CSN with an NP containing an overt cardinal determiner is well formed. Rather, (61b) is anomalous in being interpreted as one head associated with three men, while (61a) lends itself to an interpretation of one head per person more easily. Likewise, in (62b), the CSN derivation rules out multiple pictures, an interpretation licensed in the FN (62a). The ungrammaticality of (61b) and (63b), and the interpretation of (62b), are naturally explained if it is assumed that they stem from the word status of CSN. Following Word Formation, the cardinal in (61)-(62) may no longer be QRed out of the word to obtain scope over *roS*, 'head' or *tmuna*, 'picture', an operation possible in FN, as the grammaticality of (61a) and the multiple picture interpretation of (62b) show. As a result one head is associated with three people in (61b), which is anomalous, but as the association of three people with a single picture is not anomalous, the effect of CSN in (62) is that of ruling out one of two possible interpretations: the one that involves wide scope for the cardinal.

A similar account can be given for the ungrammaticality of (63b). Suppose negative polarity items such as *'iS* move in LF to adjoin, or substitute, their scope marker, *lo* in (63). The ungrammaticality of (63b), in which such movement is blocked by Word Formation in CSN is thus explained.

These contrasts strongly indicate that CSN becomes a word at a stage which affects interpretation, and hence prior to PF. (and see Borer (1988) for detailed argumentation that CSN is not lexical). In line with Baker (1988) and much subsequent work, it seems plausible that CSN is the result of incorporation: the N-head of the complement is incorporated onto the head of the CSN as a whole, and that in contrast with structures proposed by Ritter (1987,1988,1991), Siloni (in press, 1994), or Hazout (1990), in which the head of CSN and its complement do not form a morphological unit.

Returning to the list of properties in (52), note that we have now suggested not only an account for property (52D), but also for property (52B). Although the formation of a word in the CSN through incorporation does not account for the positioning of adjectives as such, it does exclude their positioning directly following the head, as is the case in FN: under the assumption that CSN formation involves head to head movement and incorporation, the placement of such modifiers directly following the head is a syntactic impossibility.

(52ii) **Properties of CSN**:

A. Definiteness Spreading
✓ B. No direct modification
C. Nested order of adjectives
✓ D. Word properties

6.3 On Incorporation and Extended Projections

Consider now the properties of incorporation as they are relevant to the structure of CSN. Very much in line with Grimshaw (1991), suppose a lexical-functional complex is an Extended Projection (ExP). In fact, using Grimshaw's intuition, let us assume that in the pair F-L, where F is a functional head and L is a lexical head, and where F-L is an Extended Projection, F and L share important categorial features. Specifically, I will also assume that while an Extended Projection may contain more than one functional head, it must contain exactly one lexical head, resulting in the licit configuration for an Extended Projection in (64a), but the illicit ones in (64b-d):

(64) a. ✓ $F_1 \approx L_1$
b. * $F^*_1 \approx L^*_1$
c. * $F_1 \approx L^*_1$
d. * F*

(where the post-symbol * indicates reiteration, ≈ indicates ExP co-membership and the common subscript notates sharing of the relevant categorial features).

Consider now a slightly more complicated situation, as depicted in (65):

(65) a. $F^m \approx (F^*) \approx L$
b. $F^{m-n} \approx (F^*) \approx L$
c. L

(65a-c) gives three schematic types of ExP. The first, (65a), is a Complete one: the (single) lexical member L is dominated by the entire set of F projections specified for its categorial type. For instance, for $L=V^m$, one may assume F^m to be C, and F* to contain T^m, Agr^o, etc.[9] (65b) is a non-maximal ExP. While some functional projections are present, the maximal one is not. For instance, again for $L=V^m$, this would be an ExP not containing C. Finally, (65c) is a bare lexical head, without any functional projections in its ExP.

Suppose now that the availability of incorporation is dependent on the ExP type from which it occurs. Suppose, specifically, that incorporation is a form of licensing non-maximal ExP.[10] If this is indeed the case, we predict that incorporation from within (65a) should be blocked altogether regardless of configuration (that is, the incorporation of the head of X^m is blocked as well), but incorporation should be possible from (65b) and (65c). More specifically, suppose that incorporation from (65b), where the ExP does have functional members, but it is non-maximal, is only possible onto F. Finally, consider incorporation from (65c), where no functional projections are present altogether. In principle, it should be possible for L to incorporate into both F and L. However, incorporation into F would only be compatible with the HMC if F dominates no lexical projections. As this situation is ruled out (of 64d), a lexical head may only incorporate into a lexical head:

(66) a. $*L \leftarrow F^m \approx (F*) \approx L$
$*F \leftarrow$

b. $*L \leftarrow F^{m-n} \approx (F*) \approx L$
✓$F \leftarrow$

c. ✓$L \leftarrow L$
$*F \leftarrow$

Consider now the applicability of the incorporation schemata in (66) to the structure of nominals, assuming CSN involves incorporation from a structure essentially as in (67):

(67) [. . . N . . . [$_X$ . . . N . . .]]
CSN-head **CSN-complement**

One direct conclusion can now be drawn: to the extent that the complement in the CSN allows the incorporation of its head, it cannot be F^m. It may be either F^{m-n} or L.

Let us then turn to the investigation of the question whether it is an L or an F^{m-n}. In considering this question, some contrasts would be helpful. Thus consider the following paradigm:

(68) a. Sarvuley xulca
sleeves shirt
'sleeves of a shirt'

b. Sarvuley xulca meSubecet
sleeves shirt plaid
'sleeves of a plaid shirt'

c. Sarvuley xulca mi-flanel
sleeves shirt from flannel
'sleeves of a flannel shirt'

d. Sarvuley SaloS xulcot
sleeves three shirts
'sleeves of three shirts'

e. Sarvuley SaloS xulcot meSubacot mi-flanel
sleeves three shirts plaid from flannel
'sleeves of three plaid flannel shirts'

(69) a. meSubac xulca
plaid shirt
'plaid shirted'

b. *meSubac xulca kxula
plaid shirt blue

c. *meSubac xulca mi-flanel
plaid shirt from flannel

d. *meSubac SaloS xulcot
plaid three shirts

(70) a. ha-kilometrim ha-'arukim beyn nahariya ve-tel-aviv mukarim li heytev
the-kilometers the-long between Nahariya and Tel-Aviv familiar to me well

b. Dan rac SloSa kilometrim 'arukim
Dan ran three kilometers long

Note that the nominal complement in CSN can be modified by an adjective, can have a cardinal associated with it, and may itself take a complement as (68) exemplifies. As such, it contrasts with Construct State Adjectives, as in (69). Here the head is an adjective, and the entire phrase is an AP. The complement of that A, a nominal itself, is, however, severely restricted. It may not be modified by an adjective (69b), it may not take a complement (69c), and it may not have a cardinal associated with it (69d). Interestingly, the complement in (69a) may take a definite article, when the entire AP exhibits definiteness agreement with a definite NP. However, in such cases the complement is still barred from itself having a complement, being modified, or being associated with a cardinal:

(71) a. ha-yeled meSubac ha-xulca
the boy plaid the shirt
'the plaid shirted boy'

b. *ha-yeled meSubac ha-xulca ha-kxula
the boy plaid the shirt the-blue

c. *ha-yeled meSubac ha-xulca mi-flanel
the boy plaid the shirt from flannel

d. *ha-yeled meSubac SloSet ha-xulcot
the boy plaid three the shirts

The definite article on the complement of the Construct State Adjective in (71) is thus clearly a concord phenomenon, rather than a functional D projection. It strongly suggests that the definite article *ha-* in Hebrew does not necessarily indicate the presence of a functional category. I return to this point below.

Returning now to the paradigm in (68)-(70), note that an intermediate case, of sorts, is attested in (70). Here, the presence vs. absence of the definite article leads to a difference in interpretation. While in (70a) the long kilometers are understood as referential (the specific kilometers separating Tel-Aviv from Nahariya), this is not the case in (70b), where three long kilometers are a measure phrase, and do not refer to any specific stretch of land.

In recent studies, Szabolsci (1989) as well as Longobardi (1994) have argued that D is the locus of reference in nominals. If this is indeed so, it stands to reason that in the absence of such reference, the measure phrase in (70b) is not, in fact, a DP. Yet, it is accompanied by a cardinal and by an adjective, suggesting that it is not just a bare N. Suppose, then, that we assume that while the nominal in (69a) is a bare N, thereby accounting for its distribution, the nominal in (70b) is a non-maximal functional projection dominating an NP. For concreteness, suppose that the nonmaximal functional projection in question is a NumP, following Ritter (1995). Suppose further that F^m for NPs is DP. As a result, the following picture emerges:

(72) a.		Dm ≈	Num_m ≈	(F*) ≈	N	**referential**
b.			Num^m ≈	(F*) ≈	N	**(measure phrase)**
c.					N	**non-referential**
(73) a.	***L** ←	D^m ≈	Num^m ≈	(F*) ≈	N	**referential**
	***F** ←					
b.	***L** ←		Num^m ≈	(F*) ≈	N	**(measure phrase)**
	✓F ←					
c.	**✓L** ←				N	**non-referential**
	***F** ←					

While (72) gives the range of ExP of nominals and their interpretation, (73) gives their incorporation possibilities. Assuming now that the complement of Construct State Adjectives is a bare N, we predict the following structure:

(74)

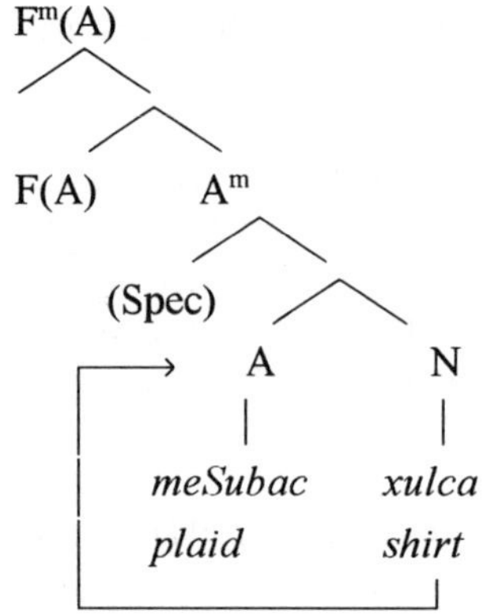

Here N incorporates onto A, remaining non-referential.[11]

Returning now to CS Nominals, recall that I assumed that the head of the complement incorporates, thereby accounting for the word properties of the CSN. It now follows from this assumption that the complement may not be a DP, as DPs would not allow further incorporation from within them. In turn, the complement of CSN is clearly not a bare N: the paradigm in (68)/(69) clearly shows that it can be modified and complemented in ways which are not available to bare N. We thus reach the conclusion that the complement in CSN is F^{m-n}, and as such, may incorporate, but only onto a functional head, not a lexical one. Having thus far proposed only NumP as an intermediate F for nominals, suppose, then, that the complement of CSN is NumP.

Consider in this view the structure of nominals. (75) is the structure of process nominals, while (76) is the structure of non-process nominals. In both, note, the head of the CSN complement, F, is now barred from incorporating onto the governing N head of the CSN. Nor is incorporation of the head of the complement onto any other head possible in these configurations, due to the HMC. Thus, a paradox seems to emerge: while the head of the complement seems to incorporate, and furthermore, it forms a word with the head N of the CSN, such incorporation appears structurally barred:

(75)

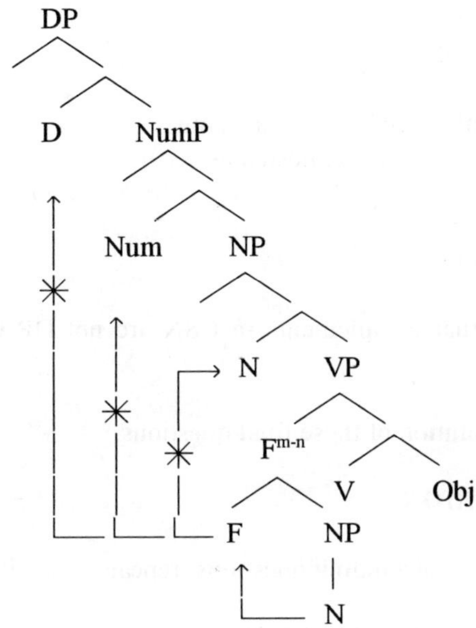

(76)

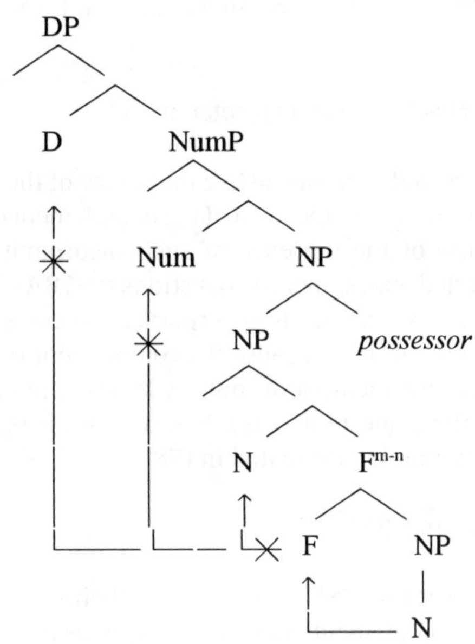

A second problem seems to face the account. I suggested that in the absence of the maximal functional projection for nominal ExP, DP, nominals are non-referential. Yet, the complement in CSN is clearly referential. It can serve as an antecedent in binding (77a), a pronominal can refer back to it (77b), and it can serve as a bound pronoun (77c):[12]

(77) a. retiyat ha-moral 'et 'acmal
view the-teacher OM herself

b. halaxnu le-beit ha-moral ve-hi$_1$ 'erxa 'otanu
went-we to-house-the-teacher$_1$ and she$_1$ hosted us

c. tmunat kalb-o$_1$ Sel kol cayar$_1$
picture dog-his of every painter

How can we reconcile the claim that complements in CSN are not DP with their referentiality?

In the next section I turn to the resolution of these final questions.

7. THE DEFINITENESS CRITERION

I concluded the last section with two outstanding questions, repeated here in (78):

(78) A. If the head of a non-maximal functional ExP can only incorporate onto a functional head, how can the N+N combination in CSN be a word formed by incorporation?

B. If the complement in CSN is F^{m-n}, why is it referential?

To these questions, a third may be added: I argued above that many of the properties of CSN can be derived from the incorporation of the head of the complement onto the head of the CSN. Further, it is because of the presence of such incorporation, and the conditions on incorporation suggested above, that the questions in (78A-B) even arise: because the head of the complement is assumed to incorporate, the complement must be non-maximal, and yet it is referential. It is because the complement is clearly not a bare lexical projection, that it cannot incorporate onto a lexical projection. In the absence of incorporation, none of these questions arise. Yet, what is the reason for such incorporation? This additional question is formulated in (78C):

(78) C. What drives incorporation in CSN?

If it is, indeed, true that the properties of CSN can follow in their entirety from the incorporation of the head of the complement, together with a restrictive theory of incorporation, the resolution of the question in (78C) becomes the key to the existence of CSN nominals altogether. Presumably, once we understand why CSN must involve incorporation, we will be able to tackle the inter-language variation question: why is it that some languages (notably Semitic and possibly Celtic languages) have CSN, while others do not?

Below, I will argue that the key to the existence and the properties of CSN is the fact that in Semitic ±definite specifications are features base generated on a lexical head.

Specifically, the account that will be proposed here for the properties of CSN does not try to derive the definiteness effects (the distribution of articles, Definiteness Spreading) from syntactic structure. Rather, the assumption is that the definiteness effects are the basic ones, and the syntactic structure is their result.

7.1 ±Definite as a Feature

In Borer (1988) I suggest that ±definite in Semitic is a feature base-generated on the head, whose value can percolate up a word structure. Some evidence that ±definite in Semitic is, indeed, a feature, comes from the presence of definiteness concord in adjectives; a particularly striking case is repeated in (79):

(79) ha-yeled meSubac ha-xulca
the boy plaid the shirt
the plaid shirted boy

The reason (79) is striking is because the agreeing adjective, *meSubac ha-xulca*, 'plaid skirted', is itself a construct, containing a nominal which carries the ±definite specification. And yet, in sharp contrast with definite nominals, it is not referential, nor does it have any of the properties typically associated with DPs. Here, the definite article on *xulca*, 'shirt', is clearly an agreeing feature, without any semantic value whatsoever.

Another striking example is given by the synonymity of (80a) and (80b), both used side by side in Modern Hebrew (although (80a) belongs to a slightly higher linguistic register):

(80) a. yalda zot tikra 'et ha-Sir
girl this will-read OM the-poem
'this girl will read the poem

b. ha-yalda ha-zot tikra 'et ha-Sir
the-girl the-this will-read OM the-poem
'this girl will read the poem'

The presence of the article on a demonstrative is clearly semantically vacuous. There is absolutely no difference in meaning between (80a) and (80b). Furthermore, if the demonstrative, at some level of representation, occupies some D position, that position is plausibly in competition with the definite article (as it is in English). This thus further suggests that the definite article *ha* is merely a feature here, base-generated on the stem.

7.2 The Definiteness Criterion

The availability of ±definite as a feature specification on a lexical head patterns with another property assumed here for Semitic: D in Semitic is unspecified for its ±definite value. Rather, it inherits it from a moved N thus specified. These assumptions are summarized in (81):

(81) a. ±definite in Semitic is a feature base-generated on the head N.
b. D in Semitic is inherently unspecified (at the base) for a ±definite value. It becomes definite as a result of a ±definite N moving into it.

I will further assume that Semitic does not have a [Spec,DP] position, and hence D cannot become ±definite as a result of Spec-Head Agreement between the head D and some nominal moving to [Spec,DP], overtly or covertly (in LF).

Suppose now that the formation of CSN is linked to the feature nature of ±definite. Specifically, suppose that the salient property of CSN is as given in (82):

(82) **The salient property of CSN**:
the N-head of CSN is base-generated without ±definite specification

Now clearly, CSN can be referential, and can be specified as ±definite. It thus follows that they are, indeed, dominated by a DP. Yet, if D in Semitic in inherently unspecified as ±definite and must inherit this feature from a moved ±definite N, a DP dominating an N generated without ±definite specification can never, itself, become ±definite.

Let us now assume, as is intuitively clear that such a situation leads to ungrammaticality. Specifically, while (83a) is a well-formed configuration, (83b) is not:

(83)
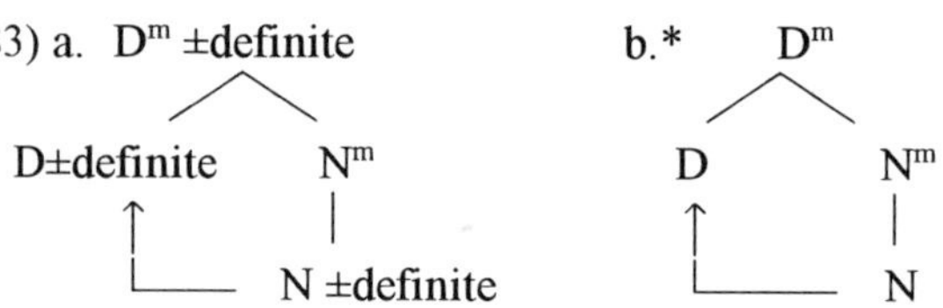

Formalizing the intuition in (83) is the Definiteness Criterion in (84):

(84) **The Definiteness Criterion**:
a. D^m must be αdefinite
b. Nαdefinite iff Dαdefinite, where N and D are members of the same ExP.

The structure in (83b) fails the Definiteness Criterion on the first account: D^m is not definite. It is thus clear that if the head of CSN is generated without ±definite specification, but is dominated by D^m, there must be a way of providing the missing ±definite specification, or a violation of The Definiteness Criterion would result. It is for this reason, I claim, that incorporation must occur in CSN. Specifically, if the head of the complement nominal is specified as Definite, its incorporation onto the head of the CSN itself would render that N Definite, and movement onto D would result in D^m being Definite as well.

Recall, now, that I argued that the head of the complement, a non-maximal ExP, cannot incorporate directly into the head N. that head N being lexical. However, the head of the complement may incorporate directly into D, providing the structural configuration allows such incorporation. Suppose, then, that the complement, a NumP, moves into a position directly governed by D. This configuration is given in (85): [13]

(85) a.

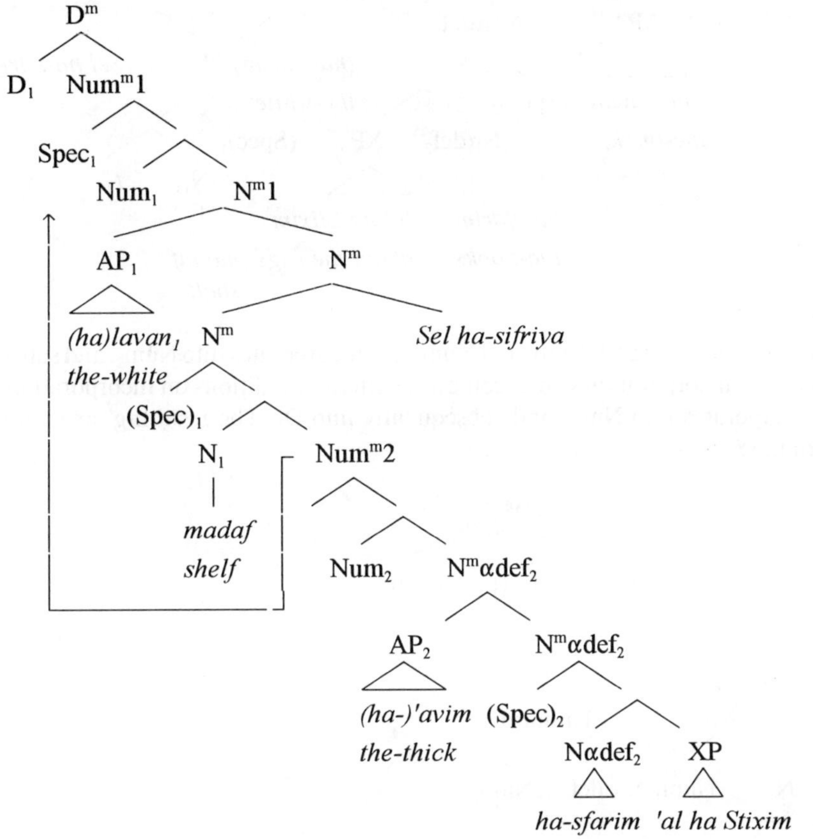

b.

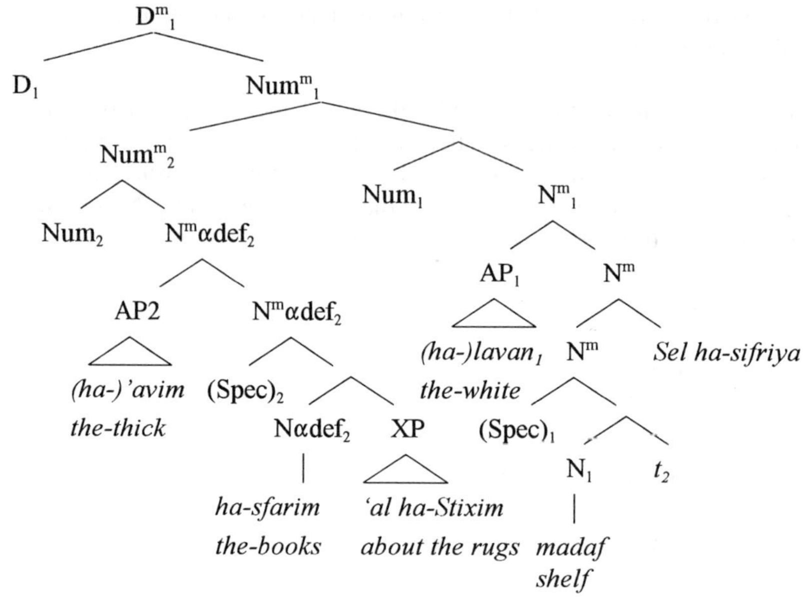

In (85b), D governs both Num_1 and Num_2. N2 incorporates into Num_2, and subsequently into D_1, an incorporation which complies with the conditions on incorporation in (66). N_1 incorporates into Num_1 and subsequently into D_1. The resulting structure of D is given in (86):

(86)

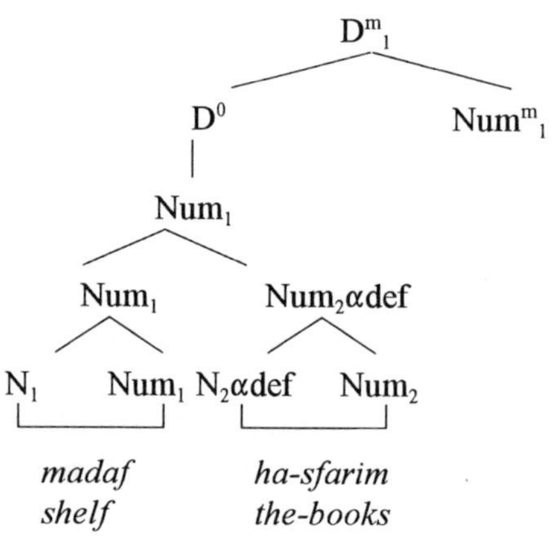

Several things should be noted with respect to (86). First, as D is empty, the movement to it is presumably substitution, rather than adjunction. As Num_1 is an ExP member of D_1, it is plausibly the head of D_1, moved by substitution, while Num_2 adjoins to it. The resulting configuration, in turn, allows Secondary Percolation (cf. Lieber, 1980; Borer, 1988): within a morphological structure, such as (87), and in the absence of some feature specification for the head, that feature may percolate from the complement. In turn, the feature associated with the root node may percolate down to the head, by a convention of feature sharing internal to Extended Projections:

(87) a.

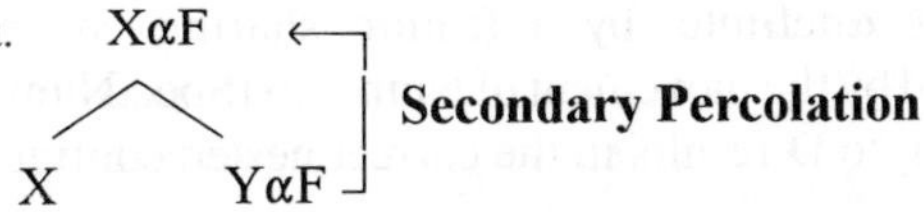

no specification for F

b. **by feature sharing internal to ExP**

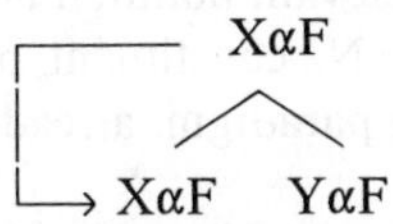

Returning to the structure in (86), the feature αdef may now percolate from Num_2 to Num_1, and in turn, to D_1, and to N_1. In turn, by feature sharing, internal to ExP, $Num^m{}_1$, the trace of Num_1, N^m, and the trace of N_1 become αdefinite as well. The resulting structure is (88):

(88)

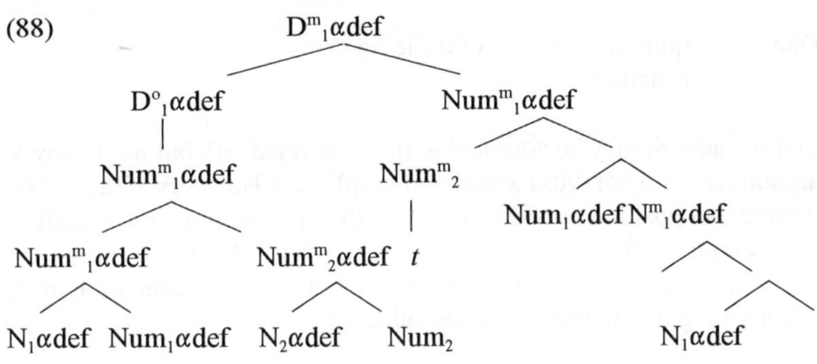

Reviewing the structure (88), it is clear that the relevant incorporation in CSN is not N_1+N_2, but rather Num_1+Num_2. This, however, clearly accords with the facts. As is well-known, both head and complement in CSN can carry number markers, both fully semantically transparent, and both triggering number agreement in modifying adjectives:[14]

(89) a. Sifrey ha-yeladot ha-'avim
book-pl.-m the-girl-pl.-f the-thick-pl.-m

b. Sifrey ha-yeladot ha-ktanot
book-pl.-m the-girl-pl.-f the-little-pl.-f

(88) is now fully compatible with the Definiteness Criterion: $D^m{}_1$ is αdefinite as required by clause (A) of the Definiteness Criterion, as a result of the incorporation of Num_2 which is αdefinite into Num_1 in D and secondary percolation $D^m{}_1$, now αdefinite, now dominates in turn an N^m which is αdefinite, by a feature sharing convention. Furthermore, the configuration formed by the movement of $Num^m{}_2$ to [Spec,Num_1], and the movement of both Num_1 and Num_2 to D results in the correct nested configuration for adjectives.

While the assumption that the movement is to [Spec,Num_1] is theory internal, empirical support for the positioning of the relevant nominal not only to the left of AP_1, but also to the right of the (incorporated) Nl±N2 constituent, presumably in D, is available.[15] In this context, consider the following paradigm, already mentioned in section 5:

(90) a. ?*ha-'axila Sel Dan be-mehirut / davka 'axSav 'et ha-tapu'ax
the-eating of Dan quickly / exactly now OM the-apple

b. 'axilat Dan be-mehirut / davka 'axSav 'et ha-tapu'ax
eating Dan quickly / exactly now OM the-apple

Assuming now that the adverb may be adjoined to the embedded VP, but not to any $\overline{V}$ projection, the ungrammaticality of (90a) was already explained. However, if the CSN--complement is fronted to a position outside the VP, as the movement to a dominating [Spec,Num^m] would entail, (90b) is predicted to be grammatical: placement of the adverb between the subject of the VP and the complement of V is still compatible with adjunction to VP, following the movement of the subject.[16]

Further evidence comes from the paradigm in (91)-(92):

(91) a. madaf ha-sfarim ha-lavan
shelf the-books the-white-sg

b. madaf ha-sfarim ha-'avim
shelf the-books the-thick-pl.

c. ?madaf ha-sfarim ha-'avim ha-lavan
shelf the-books the-thick-pl. the-white-sg

d. madaf sfarim 'avim lavan
shelf books thick-pl. white-sg

e. *madaf ha-sfarim ha-lavan ha-'avim
shelf the-books the-white-sg the-thick-pl.

(92) a. *madaf ha-sfarim ha-lavan 'al ha-Stixim
shelf the-books the white-sg about the-rugs

b. madaf ha-sfarim ha-'avim 'al ha-Stixim
shelf the-books the-thick-pl. about the rugs

c. ?madaf ha-sfarim 'al ha-Stixim ha-lavan
shelf the-books about the rugs the-white-sg

While nested AP configurations in CSN are, to begin with, cumbersome, a crossing configuration, as in (91e) is clearly much worse. It was already noted that the configuration in (85b) generates the nested configuration directly. Consider, however, the positioning of complements. Specifically, note that (92a), where the head of the CSN-complement precedes the AP modifying the entire CSN, but its complement follows it, is completely ungrammatical. Considering, now, (93)-(94), note that the CSN complement, argued here to be Num^{m}_{2}, must precede any complementation of N_1 in its entirety:

(93) a. **misgeret me-'ec** Sel tmuna Sel xamaniyot
frame from wood of picture of sunflowers

b. **misgeret** tmuna Sel xamaniyot **me-'ec**
frame picture of sunflower from wood

c. ***misgeret** tmuna **me-'ec** Sel xamaniyot
frame picture from wood of sunflowers
(only has the interpretation of a wooden picture, not a wooden frame)

(94) a. **krixa mi-karton** Sel sefer 'al Stixim
a cover from cardboard of a book about rugs

b. **krixat** sefer 'al Stixim **mi-karton**
cover book about rugs from cardboard

c. ***krixat** sefer **mi-karton** 'al Stixim
cover book from cardboard about rugs
(only has the interpretation of a cardboard book, not a cardboard cover)

Let us now turn back to the questions in (78), and see how the account given in this section for the motivation for incorporation and its nature fares with respect to them. The questions are repeated here:

(78) A. If the head of a non-maximal functional ExP can only incorporate onto a functional head, how can the N+N combination in CSN be a word formed by incorporation?

B. If the complement in CSN is F^{m-n}, why is it referential?

C. What drives incorporation in CSN?

An answer has been given to (78C): incorporation in CSN is necessary in order to provide the DP dominating the nominal with the ±definite feature specification, by assumption missing on its own N head. In turn, and given the conditions on incorporation proposed above, the complement in CSN may not be a maximal ExP, and I suggested that it is a NumP.

Recall now that a non-maximal ExP can only allow its head to incorporate onto a functional head. However, following the preposing of the NumP complement, $NumP_2$, to [Spec,Num_1], a movement clearly empirically necessary, the incorporation of Num_2 (itself an incorporated form of N_2+Num_2) may incorporate into D, which, in turn, hosts Num_1, itself an incorporated form of N_1+Num_1. Thus the incorporation attested in CSN is that of Num_1+Num_2, and it is well-formed according to the conditions on incorporation. Thus the question in (78A) is answered as well. Finally, we now account for the referentiality of the complement of CSN, although it is base generated as a NumP, rather than DP: having incorporated into a DP_1, DP_1 can now serve as D^m for both $NumP_1$ and $NumP_2$, rendering both maximal ExP, and fully referential.

8. SUMMARY

By way of summary, let us speculate briefly on the nature of inter-language variation, such that an account can be suggested for the presence, vs. absence, of CSN in grammars.

It has been proposed here that the feature nature of ±definite specification, and its base-generation on nominal stems allows for CSN formation, by allowing the stem to be base-generated without such specification. In turn, the ±definite specification on the stem patterns with an unspecified D head, requiring N movement to it in order to render it ±definite.

Consider now a language such as English. Suppose in English ±definite is not a feature, as is plausible. English does not have definiteness concord, nor is there evidence for semantically redundant use of the definite article. On the other hand, D in English is directly marked as ±definite through the existence of an independent determiner. Thus, while the structure of nominals in Semitic is as in (95a), the structure of nominals in English is as in (95b):

(95) a.

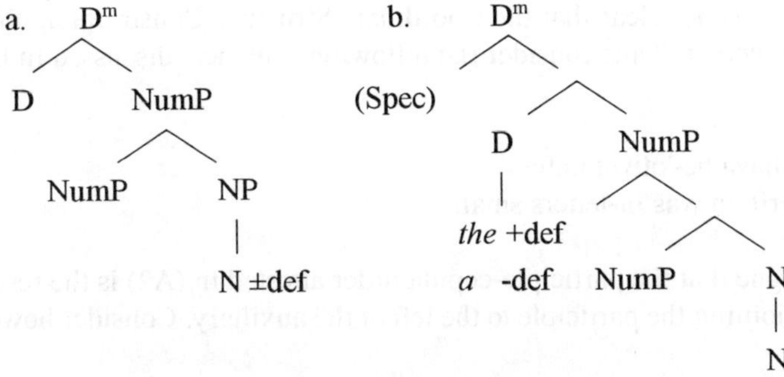

As an English D is directly marked for ±definite, and as stems are not thus marked, the configuration in Semitic, where an unmarked stem cannot transmit its properties to D, and thus requires incorporation, is not attested.

There is, however, one type of nominal in English which appears to be inherently marked as ±definite: proper names. Interestingly, Longobardi (1994) argues precisely for the N-to-D movement of such nominals in Italian, showing their syntactic position to differ from that of regular, non-inherently marked nominals. On the other hand, if the availability of ±definite as a feature on a lexical stem is in complementary distribution with the need for D to be thus specified at the base, we predict that proper names and articles should never co-occur in an English-type language, but could co-occur in Semitic. This prediction, although not perfectly correct, is clearly on the right track. Thus while Standard Arabic allows for the Definite marking of all names, English bars such marking, on the whole.[17]

A full typological study of the correlation between ±definite as a feature, the properties of determiners, and the presence of CSN nominals is clearly outside the scope of this paper. It is hoped that future research will shed more light on this correlation.

APPENDIX: THE COORDINATION PROBLEM

An interesting problem for the incorporation account for CSN (as proposed in Borer 1988, but applicable to the present work as well) is pointed out by Siloni (in press, 1994). Siloni notes that such incorporation would have to operate in violation of the Coordinate Structure Constraint, as exemplified by (A1):

(A1) beit ha-rabi mi-kiryat'arba ve-ra'ayato
house the-rabbi from-kiryat Arba and-wife-his

Recall that incorporation of a fully specified N (through a Num head) into an unspecified Num is necessary for CSN formation. In (Al), however, this N appears to be *ha-rabi* 'the rabbi', originating within the coordinated phrase *ha-rabi mi-kiryat 'arba ve-ra'ayato* 'the rabbi from Kiryat Arba and his wife'.

On the other hand, it is not clear that the Coordinate Structure Constraint applies to Head Movement altogether. Thus consider the following sentence, discussed in Borer (1995):

(A2) ha-sefer katuv haya be-'otiyot ktanot
the-book written was in-letters small

In Borer (1995) I argue that the participle-copula order attested in (A2) is the result of head movement, adjoining the participle to the left of the auxiliary. Consider however, the following:

(A3) ha-sefer katuv haya u-mudpas be-otiyot ktanot
the-book written was and-printed in letters small

In this case, it appears, participle adjunction to the copula would require movement from a coordinate head structure, as the structure in (A4) illustrates (irrelevant details omitted):

(A4) [the book [$_{FP}$ was [$_{VP}$ [$_{V}$ written and printed] in small letters]

Similar cases of head movement which are in violation of the coordinate structure constraints are discussed by Tomioka (1992) in Japanese and are attested in languages argued to have long head movement (cf. Rivero (1991), among others).

Examining, however, the structures proposed above for CSN from the perspective of the Coordinate Structure Constraint results in the exposure of a number of other problems. Thus the assumption shared by Ritter (1991), Siloni (1994) and the present work, according to which adjectives are generated to the left of the modified NP, and the surface order is derived from head movement across them runs into problems as well in such constructions, quite independent of CSN, as (A5) exemplifies:

(A5) a. ha-rabi ha-loxem ve-ha-rabanit ha-xasuda mi-kiryat'arba
the-rabbi the-militant and the rabbi's-wife the-pious from Kiryat Arba

Recall that the positioning of the adjective to the right of the head comes about as a result of the movement of the N head to D. However, as is entirely evident from (A5), only one of the conjuncts can make use of D2, while the second N head has no landing site (note that an across the board raising of N would not be helpful, as it would predict that both N would precede both adjectives, contrary to (A5)).

For Hazout (1990), who adjoins APs in Hebrew to the right of the complementation domain, and hence requires no head movement to derive their placement to the right of the head, the placement of adjectives in (A5) does not appear to present a problem, although another problem presents itself: the placement of the AP between the head and the complement PP can only follow, given rightward adjunction to $\overline{\text{N}}$, if PP is post-posed. Such a PP in (A5) would have to be post-posed from a coordinated structure. Yet another coordination problem plagues the Hazout account. Thus consider the sentence in (A6):

(A6)dirat ha-rav ha-yafa u-migraSav ha-rabim ba-'ir
apartment the-rabbi the-pretty and lots the-many in-the-city
(problematic for Hazout (1990))

According to Hazout, in (A6) *dirat*, 'apartment', must move to D for CSN to be formed. However, such movement would clearly be from a coordinated structure, and not a coordination of heads, either.

As clearly all accounts of CSN and the distribution of adjectives in Semitic are susceptible to problems with the CSC, two possibilities are available for a solution. Possibly, attempting to shed light on the structure of nominals in Semitic using head movement is misguided. Another possibility, however, is that the CSC, a rather poorly understood constraint to begin with, is suspended in these cases for reasons that must require additional explanation. One clear solution to these cases (which applies to the proposal advanced in this paper, as well as to the proposals advanced in Siloni (1994) and Ritter (1991) but is not helpful as far as Hazout (1990) is concerned) is to argue that coordinated nominals are always DPs, and that in cases such as (A5), in which a single PP modifies both conjuncts involves ellipsis or adjunction to the coordinated DP.

NOTES

* Various aspects of this research have been presented in the Afroasiatic Conference in Sophia Antipolis and in Tel-Aviv University, and have been incorporated into graduate seminars in the Hebrew University and at the University of Massachusetts. I thank audiences in these places for insightful comments. Special thanks go to Betsy Ritter, Ur Shlonsky, and Tali Siloni for ongoing discussions on the structure of Hebrew nominals, and to Kyle Johnson for his editorial interventions.

[1] (25a) is actually ambiguous between the reading given (*successful* having scope over *second*), and the reading assigned to (25b) (*second* having scope over *successful*). No such ambiguity is associated with (25b). This ambiguity seems to suggest that possibility of post-posing from the structure in (25b) to create the structure in (25a). I leave this matter to future research.

[2] See, especially, Siloni (in press), (1994), as well as counter-arguments in Borer (forthcoming).

[3] Shlonsky further argues that in the presence of a complement, an *agent* and a *possessor*, all three constituents can occur in any order. In my dialect, however, some combinations, notably head-*possessor*-*agent*-complement, head-*agent*-*possessor*-complement, are strongly dispreferred. The relative order of *agent* and *possessor* in NPs or its interaction with the order of complements, are largely irrelevant to this study, and in the examples below I will concentrate on the contrast between the distribution of the complement, on the one hand, and both *possessor* and *agent* on the other hand.

[4] See Hazout (1990) for arguments against a passive in VP analysis for (39). See Borer (1991, forthcoming) for counter-arguments.

[5] The impossibility of CSN formation with a (right-)adjoined DP predicts that FN may appear freely with a *possessor/agent* but no complement, but CSN of this form should be ungrammatical. The grammaticality of FN construction is illustratate in (i). CSN with a *possessor/agent*, on the other hand, are highly restricted, but at times, for poorly understood reasons, not fully ungrammatical:

(i) a. ha-tmuna Sel van gogh
the-picture of Van Gogh

b. ha-tmuna Sel ha-muzeon
the-picture of the museum

(ii) a. ?*tmunat van gogh
pictures Van Gogh

b. *tmunat ha-muzeon
picture the-museum

Rosen (1956) argues that *head+possessor* CSN is only possible if it indicates part-whole relations, plausibly complementation rather than *possessor*-head relationship, or inherent possession. The picture is more complicated, however. Thus plurality of the head N improves (although not beyond marginality) the ungrammaticality of (iia), but not of (ib). This, and related matters, are left for future discussion.

[6] This fact, note, is accounted for quite independently of whether a VP is involved in process nominals. Proponents of Complex Event Nominals (in the sense of Grimshaw, 1990) would only need assume that an argument structure is involved. In either case, realization of the subject without the object would be blocked.

[7] Note that it is not possible to assume that the *possessor* in non-process nominals is sometimes an adjunct, accounting for NOS orders, and sometimes in [Spec,NP] accounting for NSO orders, as the latter configuration would erroneously allow the subject to move to [Spec,FP] contrary to fact, allowing the formation of a CSN.

For the sake of completeness, one must exclude the possibility that *possessor/agents* in non-process nominals may be adjoined both to the left and to the right, thereby making object post-posing unnecessary. Some evidence that left adjunction is, indeed, blocked, comes from cases where a *possessor* is present alongside argumental subjects, as in (i). While (ia-d) might be cumbersome for some speakers, they are not ungrammatical, and they contrast sharply with the ungrammatical cases in (iia-d):

(i) a. hafakat hamlet 'al yedoy Branagh Sel teatron ha-renesans
production Hamlet by Branagh of the Renaissance Theater

b. hafakat Branagh 'et hamlet Sel teatron ha-renesans
production Branagh OM Hamlet of the Renaissance Theater

c. ha-hafaka Sel hamlet 'al yedey Branagh Sel teatron ha-renesans
the-production of Hamlet by Branagh of the Renaissance Theater

d. ha-hafaka Sel Branagh 'et hamlet Sel teatron ha-renesans
the-production of Branagh OM Hamlet of the Renaissance Theater

(ii) a. *hafakat hamlet Sel teatron ha-renesans 'al yedey Branagh
production Hamlet of the Renaissance Theater by Branagh

b. * hafakat Branagh Se l teatron ha-renesans 'et ham let
production Branagh of the Renaissance Theater OM Hamlet

c. *ha-hafaka Sel teatron ha-renesans Sel hamlet 'al yedey Branagh
the-production of the Renaissance Theater of Hamlet by Branagh

d. *ha-hafaka Sel teatron ha-renesans Sel Branagh 'et hamlet
the-production of the Renaissance Theater of Branagh OM Hamlet

If, indeed, left adjunction of the *possessor/agent* were possible, we would predict the grammaticality of (iia-d) to be exactly on a par with that of (ia-d). That this is not so indicates that left adjunction is excluded.

[8] A problem with Hazout's (1990) right adjunction of adjectives which is not addressed here in detail is the fact that his structure erroneously predicts the grammaticality of (24b).

[9] A hidden assumption connected with the configurations in (65)-(66) must be made explicit: I assume that non-maximal ExP may only be missing functional projections from the top down, thus, assuming the maximal F projection for V to be C, Agr° ≈ V is a possible non-maximal ExP, but C ≈ V is not.

[10] For a similar idea, see Koopman, (1994), where it is argued that non-maximal lexical projection may satisfy their functional selection matrix through incorporating onto other lexical projections.

[11] Interestingly, Construct State Adjectives could function as referential nominals, as in the following

(i) meSubac ha-xulca higi'a.
plaid the shirt arrived
'The plaid shirt arrived.'

In (i), *meSubac ha-xulca* 'the plaid shirt' is clearly not an agreeing adjective, but a referential nominal, serving as the subject argument of *higita* 'arrive'. Plausibly, in this case the AP is dominated by a DP, rendering it referential. The option of being dominated by a DP, thereby becoming referential, is available for all adjectives in MH, as it is, peripherally, for English (Cf. *The Good, the Bad and the Ugly)*. In fact, as noted by Longobardi (1994), it is only in the presence of an overt definite article that such adjectives may be used as referring expressions. When AP is not dominated by a DP, and hence not a referring expression, it is, presumably, dominated by some functional projection unique to modifiers, possibly a degree phrase. These issues are not pursued here any further.

[12] The binding properties of complements in CSN may appear, at first sight, to contradict the claim that the CSN is a word. Recall that bound pronouns could not be licensed from within CSN. The difference is, however, that in the latter case QR was necessary, and it is precisely that operation which is blocked from inside a word. This, I assume, is not the case for anaphoric binding.

[13] For the sake of simplicity, the functional projection dominating AP is ignored in these structures.

[14] The situation depicted in (89) contrasts with the situation in compounds, which share many of the properties of CSN, but where the complement of CSN is never semantically transparent, and may never be modified by an adjective or take a complement (see Borer (1988) for discussion). Likewise, in Construct State Adjectives, the complement N may be plural, as in *'adom re'afm*, 'red slated' literally red slates, but the plurality of *re'aSm 'slates'* is semantically vacuous here, cannot be modified and cannot take a complement. Further, Hebrew allows us to express an identical meaning with the singular, *'adom ra 'at* literally red slate although there may be numerous slates.

[15] Recall that the movement of the relevant nominal constituent in CSN, but not in FN, over the adjoined adjective is argued for by Ritter (1991) and Siloni (1994) as well.

[16] A general problem concerns the impossibility of the following word orders in both sentential VPs and VPs embedded in process nominals::

(i) *XP adv subj V obj
*la-yeladim ba-xaSay Rina Salxa matanot
to-the-boys secretly Rina sent presents

*XP V adv subj obj
*la-yeladim Salxa ba-xaSai Rina matanot
to-the-boys sent secretly Rina presents

(ii) N±V adv subj t obj
*ha-miSlo'ax ba-xaSay Sel Rina 'et ha-matanot
the-sending secretely of Rina OM the-presents

While the parallelism between VPs in sentences and in process nominals is maintained here, within the VP-internal hypothesis no direct account is available for the ungrammaticality of (i)-(ii).

[17] Well-known exceptions to this generalization are proper names such as *The Bronx* in English, or *La France*, in French. For some interesting discussion which is couched in terms rather similar to those proposed in this paper see Longobardi (1994).

REFERENCES

Baker, M.: 1988, *Incorporation: a theory of grammatical function changing*, Chicago: Chicago University Press.

Borer, H.: 1984, *Parametric Syntax*, Foris Publications, Dordrecht.

Borer, H.: 1988, 'Morphological Parallelism: a Case Study', In G. Booij, ed., *Morphology Yearbook* 1, Foris Publications, Dordrecht.

Borer, H.: 1991, 'The syntax and morphology of derived nominals', ms. University of Massachusetts.

Borer, H.: forthcoming, *Parallel Morphology*, to be published by MIT Press.

Borer, H.: 1995, 'The Ups and Downs of Hebrew Verb Movement', *Natural Language and Linguistic Theory*, 13, 527-606.

Chomsky, N.: 1992, 'A Minimalist Program in Linguistic Theory', *MIT Occasional Papers in Linguistics* 1, Cambridge, Mass.

Duffield, N.: 1992, 'Construct State in Irish and Hebrew', ms. Heinrich Heine Universität, Dusseldorf

Emonds, J.: 1978, 'The Verbal Complex V'-V in French', *Linguistic Inquiry* 9, 49-77.

Fassi Fehri, A.: 1988, 'Case Inflection, VS Order and X'-Theory', In Fassi Fehri, ed., *Proceedings of the 1st International Conference of the Linguistic Society of Morocco.*

Fassi Fehri, A.: 1989, 'Generalized IP Structure, Case and VS Word Order', Iin I. Laka and A. Mahajan, eds., *MIT Working Papers in Linguistics* 10, 75-113.

Fassi Fehri, A.: 1993, *Issues in the Structure of Arabic Clauses and Words*, Kluwer Academic Publishers, Dordrecht, The Netherlands.

Grimshaw, J.: 1990, *Argument Structure*, MIT Press, Cambridge, Mass.

Grimshaw, J.: 1991, 'Extended Projections', ms. Rutgers University

Hazout, I.: 1988, 'Hebrew derived nominals', ms., University of Massachusetts, Amherst.

Hazout, I.: 1990, *Verbal Nouns: Theta-Theoretical Studies in Hebrew and Arabic*, Doctoral Dissertation, UMass Amherst.

Hazout, I.: 1995, 'Action nominalization and the Lexicalist Hypothesis', *Natural Language and Linguistic Theory*, 13, 355-404.

Koopman, H.: 1994, 'Licensing Heads', ms., UCLA.

Lieber, R.: 1980, *On the Organization of the Lexicon*, Doctoral Dissertation, MIT, Cambridge.

Longobardi, G.: 1994, 'Reference and Proper Names', *Linguistic Inquiry*, 25, 609-666.

Muhammad, M.: 1988, 'On the Parallelism between IP and DP', *Proceedings of WCCFL* 7, 241-254.

Ouhalla, J.: 1990, *The Syntax of Head Movement: a study of Berber*, Doctoral Dissertation, University College of London.

Ouhalla, J.: 1991, *Functional Categories and Parametric Variation*, Rutledge, London.

Prince, A.: 1973, Doctoral Dissertation, MIT, Cambridge, Mass.

Ritter, E.: 1987, 'NSO orders in Modern Hebrew', J. McDonough and B. Plunkett, eds., *Proceedings of NELS* 17:521-537.

Ritter, E.:1988, 'A Head Movement Approach to Construct-State Noun Phrases', *Linguistics* 26, 909-929.

Ritter, E.: 1991, 'Two Functional Categories in Noun Phrases: Evidence from Modern Hebrew', *Syntax and Semantics* 25, 37-62.

Ritter, E.: 1995, 'On the Syntactic Category of Pronouns and Agreement', *Natural Language and Linguistic Theory* 13, 405-443.

Rosen, H.: 1956, *Ivrit Tova* ('Good Hebrew'), Kiryat Sefer Publishers, Jerusalem.

Szabolsci, A.: 1989, 'Noun Phrases and Clauses: Is DP analogous to IP or CP?' In J. Payne, ed., *Proceedings of the Colloquium on Noun Phrase Structure, Empirical Approaches to Language Typology*, Mouton, Den Hague.

Shlonsky, U.: 1988, 'Government and Binding in Hebrew Nominals', *Linguistics* 26, 951-976.

Siloni, T.: in press, 'Hebrew Noun Phrases: Generalized Noun Raising', A. Belletti and L. Rizzi, eds., *Issues in Comparative Syntax*, Oxford University Press, Oxford.

Siloni T.: 1994, *Noun Phrases and Nominalization,* Doctoral Dissertation, University of Geneva, Geneva.

Tomioka, S.: 1992, 'Verb movement and tense specification in Japanese', In J. Mead, ed., *Proceedings of WCCFL* 11.

Valois, D.: 1991, *The internal syntax of DP*, Doctoral Dissertation, UCLA.

THE INTERNAL AND EXTERNAL DISTRIBUTION OF PRONOMINAL DPs*

Hilda Koopman
University of California, Los Angeles

The form and distribution of pronouns varies considerably cross-linguistically. In this paper, I will propose that there is a direct relation between the form (i.e. DP internal realization), and syntactic distribution (i.e. DP external realization).

Pronominal elements can be homophonous with determiners (French *le/la/les*, Dutch *'t*, ...), or not (English *him/her/them*, French *lui*, Dutch *'m* ...), or they can be composed, as demonstrative pronouns in many African languages, or the Welsh reduplicated pronouns discussed below). Pronominal elements can be overt or covert (e.g. *pro*). Overt pronominal elements can distribute as DPs (i.e. their distribution is identical to that of (specific) DPs), as clitics (their distribution is that of a head which occurs in a particular syntactic position), or as agreement morphology (i.e. as a head that shows up if an XP is or has moved through its Spec position).

The analysis of the syntax of pronominal elements meets with considerable analytical difficulties. Whether a particular pronominal element has the distribution of a DP or not, presupposes an understanding of the distribution of the relevant type of DP. Whether distributional differences can be accounted for by a combination of the distribution of DPs, and heads (i.e. whether a pronoun is a head that has undergone head movement from the DP position) further presupposes an understanding of the distribution of heads. Further questions arise beyond these analytical problems. If the pronominal element is a head, what position is it cliticized to? What distinguishes it from agreement? Why do pronouns, but not DPs, trigger obligatory agreement in some languages? Why do clitics and agreement occur in the particular positions they do? How should pro-drop be analyzed? How are clitic doubling constructions to be analyzed? How should the often different distribution of first and second person pronouns versus third person pronouns be accounted for?

Most studies of pronominal elements base their analyses on their distribution in syntactic structures, and pay little or no attention to the actual form of the pronoun.[1] This may be because there is not much overt evidence on which to base an analysis, or because there does not seem to be much at stake, or simply because the internal structure of DPs has not been developed enough to allow this question to be asked. I argue in this paper that a proper understanding of the form and structure of pronominal elements is in fact extremely important: there is an intimate connection between the DP internal syntax and the DP external syntax.

K. Johnson and I. Roberts (eds.), Beyond Principles and Parameters. 91-132.

I will start out with an analysis of the internal structure of pronominal DPs, and show that pronominal DPs consist of (at least) two basic elements, NumP and D,[2] which can be in an overt agreement relation, with NumP in Spec, DP. Languages vary with respect to overtness or covertness of NumP and of D. This defines a typology of possible pronouns, and has immediate consequences for the external distribution of pronouns. In particular, the structure of pronominal DPs, taken together with movement theory, predicts the existence of NumP movement with D-stranding. NumP movement, as I will show, is the key that leads to understanding many aspects of the distribution of pronominal elements. Since NumP movement plays such an important part in accounting for the syntax of pronominal elements, I motivate its existence carefully. I establish NumP movement and D stranding on the basis of the distribution of pronouns, agreement, and lexical DPs in Welsh, which has a sufficiently rich pronominal syntax to make the empirical argument for NumP movement and D-stranding. The NumP movement analysis straightforwardly accounts for the distribution of pronouns in Mainland Scandinavian (section 6); and can be extended to English (section 7). The analysis proposed in this paper allows to reduce the important superficial differences of languages like Welsh and English to identical structures, with differences in what gets lexicalized.

1. THE FORM OF THE PRONOUN

Pronouns are specific DPs, and should therefore be represented as specific DPs. Following Postal (1969), it is often assumed that pronouns are intransitive Ds (DP --> D) (Abney (1987), Cardinalletti (1993), among others). This proposal explains the homophony of pronouns and Ds in some languages, but it fails to capture the fact that pronouns in some sense stands for nouns, encode person and number features, nor does it make explicit what kind of relation there is between between D and N. This proposal furthermore cannot accommodate pronominal forms that are not homophonous with Ds, nor compound pronouns, and merely treats the form of the pronoun and the relation to the structure as accidental.

There is no reason why pronouns should not fit into the regular DP structure. This is, in fact, the null hypothesis. Recent work on the internal structure of DPs (Abney (1987), Carstens (1991), Ritter (1991), Szabolcsi (1987), Stowell (1991), Valois (1991), among others) has shown that DPs are highly structured projections, similar to clausal projections, with at least the following structure (NumP stands for Number Phrase; AgrP, I assume, is related to the presence of a DP internal subject (i.e. a genitive DP) and is only present when a genitive DP needs to be licenced. I therefore omit AgrP from DPs not containing a genitive DP)):

(1) $[_{DP}$ Spec [D $_{AgrP}$ Agr [$_{NumP}$ Spec [Num $[_{NP}$]]]]

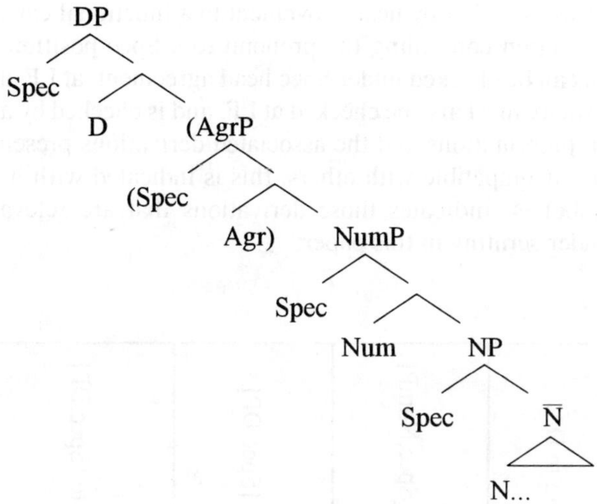

The linear order of constituents in the DP is derived by various movements and parallels the way in which the linear order in a clause is derived. Head movement accounts for displaced heads, and DP movement for the position of the structurally Case marked DP (genitive) in the overt syntax. As I show below, there are additional DP movements, of constituents *smaller* than DP.

Since pronouns are DPs (1) defines the space for cross-linguistic variation. Pronouns are specific, they stand for a head noun, and they encode person, number and gender features. We therefore minimally expect the structure in (2):[3]

(2)

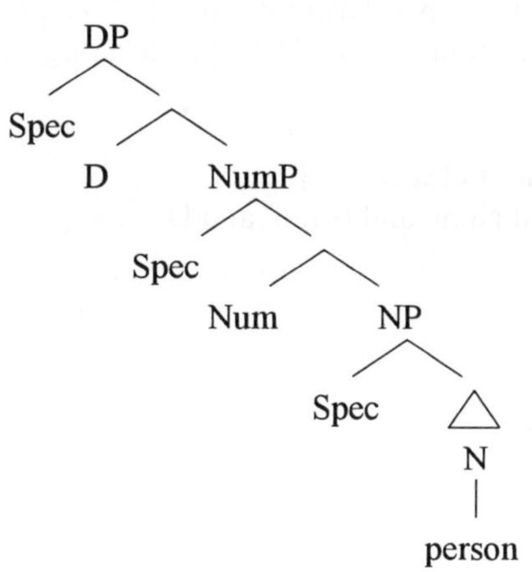

It is reasonable to assume that pronouns are lexically marked for number, person and gender features. These features must therefore be checked. In principle this can be achieved in one of two ways: either by head movement to a functional category, or by movement of some projection containing the pronoun to a Spec position where the features of the pronoun can be checked under Spec head agreement, at LF at the latest. If we assume that specificity must also be checked at LF, and is checked by a particular D, we expect the LF representations and the associated derivations presented in (3). Certain movements are incompatible with others: this is indicated with a star in the table below.[4] The symbol ☞ indicates those derivations that are relevant for the particular pronouns under scrutiny in this paper:

(3)

	N to Num	Num to D	NP to [Spec,NumP]	NP to [Spec,DP]	NumP to [Spec,DP]
a.	☞ +	-	*	*	☞ +
b.	☞ +	+	*	*	*
c.	-	-	☞ +	?	☞ +
d.	-	+	+	+	*

I should point out that, for the purposes of this paper, head movement of the pronoun for checking purposes (3b) will only play a limited role. The Spec head agreement relation will do most of the work.

(4) NP movement to [Spec, NumP]; NumP to [Spec, DP];
Spec head agreement between NP and Num, and NumP and D.

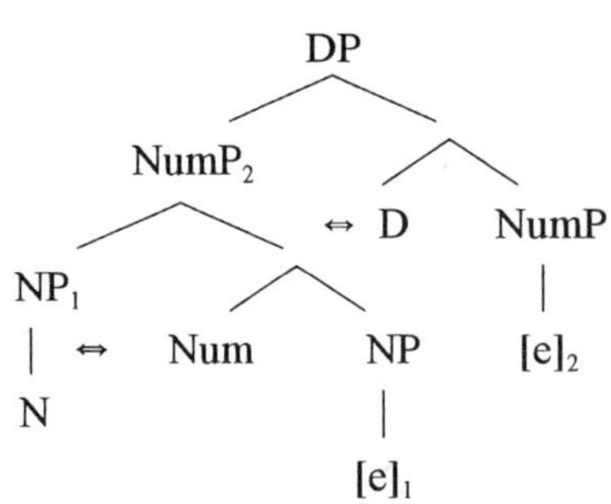

(5) N movement to Num, NumP movement to [Spec, DP].

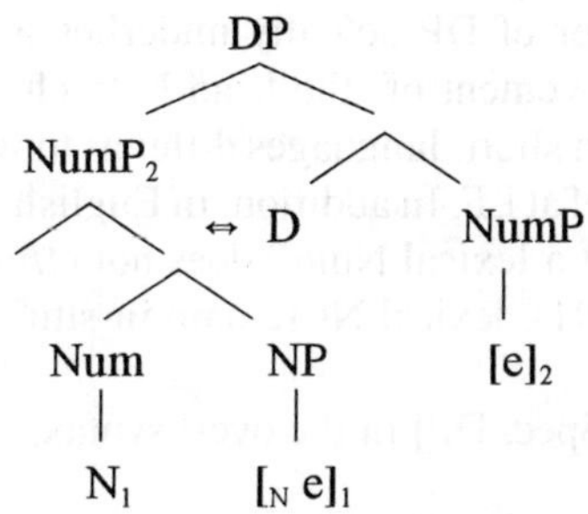

For reasons of simplicity, I abstract away from the internal structure of NumP in the remainder of this article, and simplify the structure to (6):

(6)

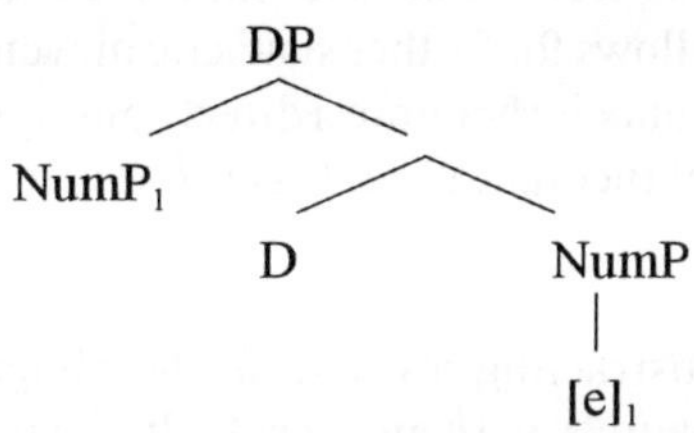

The overt syntax can reflect the various stages between (2) and (4) and (5) depending on whether movement is overt or not, how much movement is overt and what elements are lexical.

Given the background so far, we can now turn to various pronominal systems. In English, I assume, and motivate below, NP must move to [Spec, NumP], and NumP must move to [Spec, DP] in the overt syntax.[5, 6] NumP agrees with an empty D. The obligatory absence of D parallels the obligatory absence of C when [Spec, CP] is filled (cf. Koopman (1991), Sportiche (1995)).

(7) English (overt syntax):

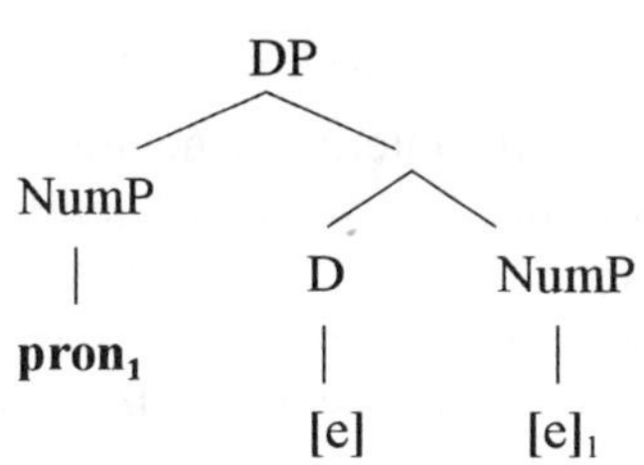

In this structure, NumP, an element *smaller* than DP, undergoes movement to [Spec, DP]. Justification for the existence of NumP movement comes from other languages. Nkmenji (1993), for example, provides quite convincing arguments for overt

NumP based on the derivation of DP internal word orders and agreement patterns within the DP for Nweh, a Grassfield Bantu spoken in Cameroon.[7] I argue in Koopman (1994) that NumP movement to Specifier of DP actually underlies a necessary LF representation which permits further movement of the head N to check selectional restrictions with the theta-role assigner. In short, languages differ as to whether NumP movement takes place in the overt syntax of at LF. In addition, in English, a pronominal NumP must move in the overt syntax, but a lexical NumP does not *(*house John the; John house the*). As shown in Valois (1991), lexical Ns remain in-situ[8]:

(8) a. pronominal NumPs move to [Spec, DP] in the overt syntax;
    b. lexical NumPs move at LF

This asymmetry is ultimately responsible for the different distributional properties of pronouns and lexical DPs. Pronominal NumP are governed by an external governor. This either allows further NumP movement (4 , 5, and 6), or head movement of the pronoun. Lexical NumP are not in a position that allows for further syntactic movement. The lack of (lexical) N incorporation in English syntax is therefore a direct consequence of the overt syntax: a lexical N cannot undergo N incorporation, because it is too low in the structure.[9]

French pronominal DPs mirror English DPs (abstracting away from the obligatory cliticization of D): a *covert* (i.e small NumP *pro*) agrees with an *overt* D. It is typically the case in French that the head is overt, and the element in the Spec silent: wh-operators in French for example are silent and the head C is overt whenever possible (i.e., when "recoverable").

(9) French:

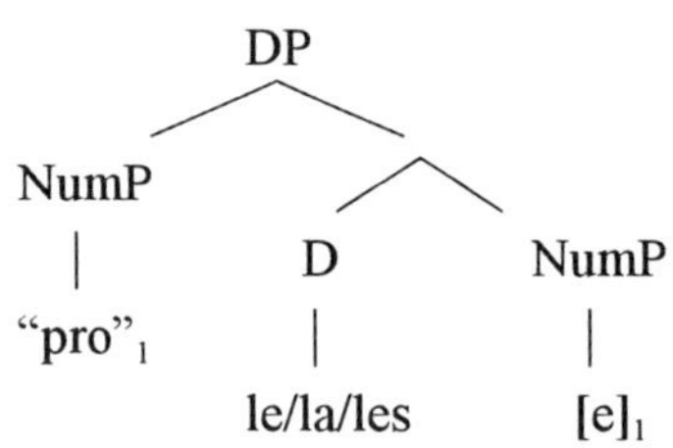

Pronominal DPs thus far differ with respect to the internal structure of NumP, whether NumP moves to [Spec, DP] in the syntax or not, and whether NumP in [Spec, DP] is overt and D is overt or covert. Restricting attention to languages with overt NumP movement to [Spec, DP], other combinations of overt/covert are expected to occur. For example, a pronominal DP with both NumP and D silent is a likely candidate for the representation of *pro*:

(10) small *pro*

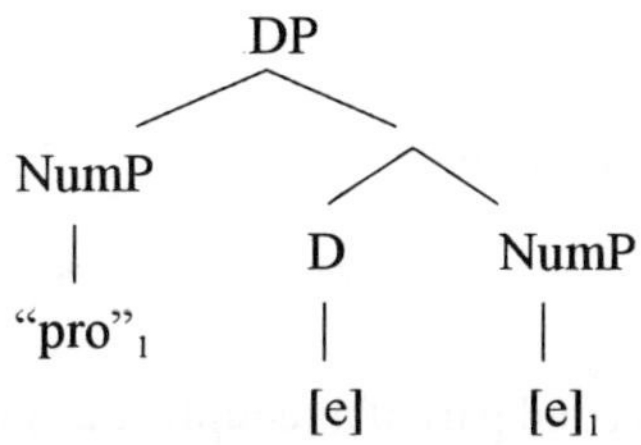

Small *pro* is further constrained by the need to be identified: in (10) covert NumP *pro* is not identified, because of the absence of overt features; identification will depend on agreement with some external head. We also expect a pronominal DP with both NumP and D overt. Welsh reduplicated pronouns fit into this slot, as I will show in the next sections.[10]

2. THE PRONOMINAL SYSTEM OF WELSH

In the pronominal system of Welsh all possibilities in terms of overt/covert NumP overt and overt/covert D seem to be attested, as I will now show. Welsh pronominal paradigms are quite complex.[11] There is a basic distinction between pronouns that can only be used in the absence of agreement (*independent pronouns*) and pronouns that cooccur with agreement (*dependent* or *auxiliary pronouns*).

I start the analysis of the Welsh pronominal system with the paradigm of reduplicated pronouns, which belong to the series of so-called independent pronouns.[12, 13]

(11) Reduplicated pronouns (Williams (1980, p.46))

1.	my*fi*	ny*ni*
2.	ty*di*	chwy*chwi*
3.	(efe, *efo* *fe/fo* (masc.)) hy*hi* (fem)	hwynt*hwy*

Third person masculine pronouns are between parentheses; they are identical to the auxiliary or affixed pronouns discussed below (except for *efe* which only occurs as a reduplicated pronoun[14]). The second part of the reduplicated pronoun is italicized to draw attention to the (significant) homophony of this part and the so-called auxiliary pronouns presented in (12). Auxiliary or affixed pronouns cooccur with agreement.

(12) Auxiliary or affixed pronouns (Williams (1980, p.48))

1.	*i/fi*	*ni*
2.	*di*,ti	*chwi*
3.	(masc.)*ef*/efo	*hwy*,hwynt
	fo/fe	
	(fem.) *hi*	

As before, auxiliary forms which occur as the second part of a reduplicated pronoun are in italics.

Reduplicated pronouns are thus clearly composed of some pronominal element followed by an *auxiliary* or *affixed* pronoun which also encodes person, number and gender as well.

The similarity between auxiliary pronouns and the second part of reduplicated pronouns in fact seems to extend beyond their segmental forms. Phonology also seems to treat them in similar ways. The italicized forms in the reduplicated paradigm are accented, which is unusual since the accent usually falls on the penultimate syllable. Auxiliary pronouns can be accented as well.[15]

Suppose that the first part of the reduplicated pronoun is a NumP, and that the auxiliary pronoun is (some kind of) a D agreeing with the NumP. This would allow auxiliary pronouns and reduplicated pronouns to immediately directly fit into the proposed structure for pronominal DPs, and describe their difference. They differ only with respect to overtness of covertness of NumP.[16]

(13) Welsh:

a. auxiliary pronouns

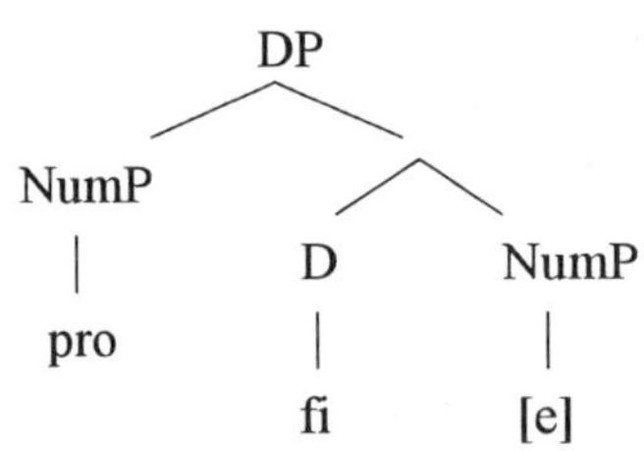

b. reduplicated pronouns

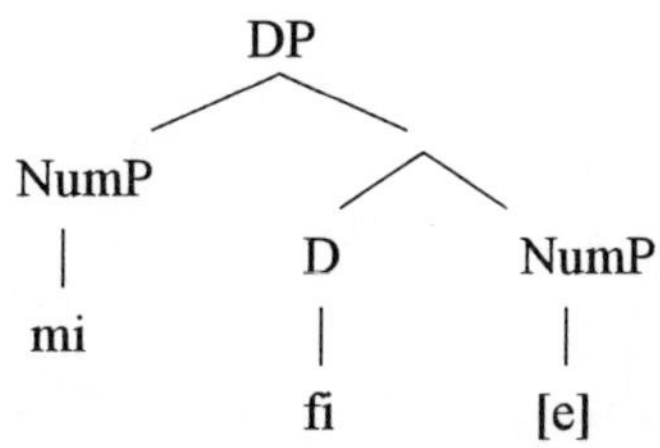

Third person pronouns (non-feminine) do not appear to allow for an overt NumP in [Spec, DP]. Some other forms in the auxiliary pronouns do not occur as reduplicated pronouns (*i, ti, hwynt*). The use of both *i* and *ti* appears to be phonologically conditioned (Williams, p. 50, 74(i)). The occurrence of *hwynt* as an auxiliary pronoun is potentially problematic, but not attested in any of the other grammars I have consulted. I will ignore this form in the remainder of this paper.

The first part of the reduplicated pronoun should thus be analyzed as NumP. At this point, a comparison between the reduplicated independent (11), the ("short") independent (14) and auxiliary pronouns (12) becomes important (as before, the forms in italics are auxiliary pronouns, i.e. the forms we have analyzed as agreeing Ds):

(14) Simple independent personal pronouns:

1.	mi,*fi*	*ni*
2.	ti,*di*	*chwi*
3.	ef (*masc.*)	*hwy*, hwynt
	hi (*fem.*)	

The non-italicized elements *my*, *mi* and *ty, ti*, and *hwynt* can unambiguously be analyzed as NumPs, since they also appear as the first part of the reduplicated pronoun, i.e. as the part that we analyzed as NumP. Welsh independent pronouns thus either have a English-like structure, with D covert and NumP overt, or a French like structure, with an agreeing overt D, and silent NumP:

(15) Independent pronouns (1st person):

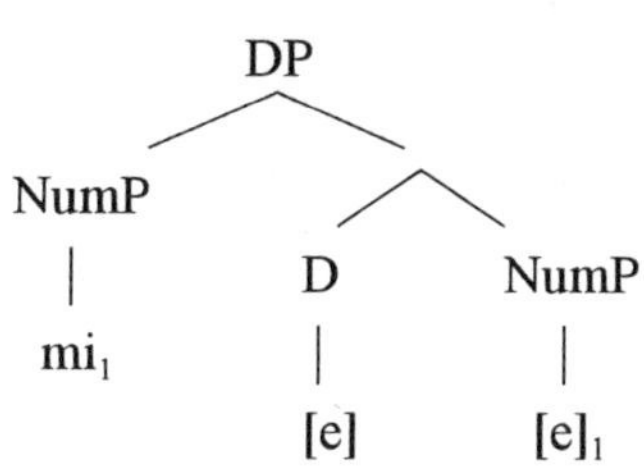

or:

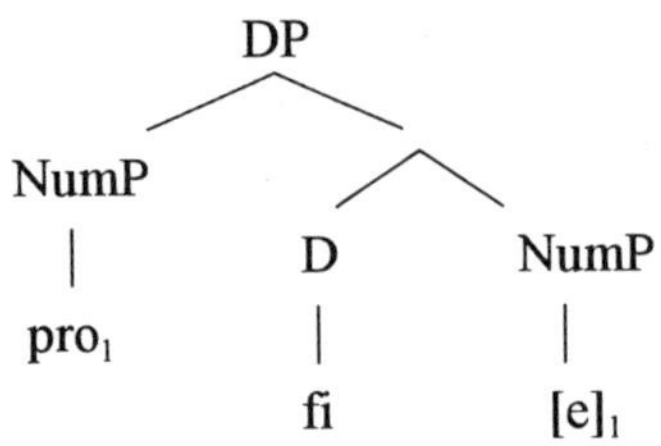

Besides these forms, there is another series of independent pronouns, called conjunctive[17] pronouns (16):

(16)

Singular	Plural
1. minnau	ninnau
2. tithau	chwithau
3. yntau (masc.)	hwythau, hwyntau

These pronouns are important in as far as they appear to be composed of an overt NumP, as can be concluded from 1s and 2s forms (*mi/ti*, not *fi, di*) and some other head (*thau*).

Finally, since Welsh also allows *pro* in agreement configurations, there are pronominal DPs with both NumP and D silent:

(17)

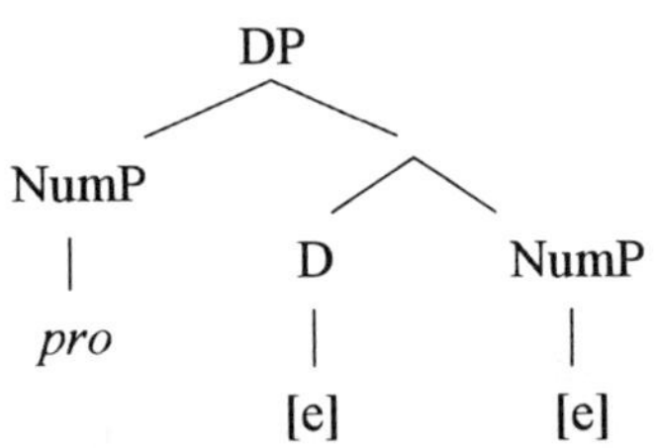

In sum, Welsh pronouns are composed of NumP and D, and all combinations of covert/overt NumP and covert/overt D are attested:

(18) *Welsh Pronominal Paradigms*:

NumP	*D*	*+ = overt, - = silent*
+	-	: some independent pronouns (e.g. *mi*)
+	+	: reduplicated pronouns, conjunctive pronouns (with special D)
-	+	: auxiliary pronouns
-	-	: *pro*

This analysis of the internal structure of Welsh pronouns becomes extremely important in the following sections, where I will propose an account of overt agreement in Welsh, based on the idea that NumP in [Spec, DP] can undergo further movement to [Spec, Agr], and thus trigger agreement, stranding D, causing apparent clitic doubling when D is overt.

3. EXTERNAL SYNTAX

Given the structure of pronominal DPs, there is a certain number of possibilities with respect to their external (i.e. syntactic) distribution. The entire pronominal DP could undergo further movement to a DP position (19a); The head of NumP in [Spec, DP] or D itself can undergo head movement when governed by an appropriate host (19b). A final possibility, given that NumP occurs in [Spec, DP], and Specifier positions are typically extraction sites, would be for NumP to move from [Spec, DP] to a further (designated) position, stranding D. The head of this position must be able to license NumP, a constituent that is smaller than a DP. I will assume that Agr, or D (as Sportiche's (1995) clitic projection) is such a position (19c):

(19) a. *DP movement*:

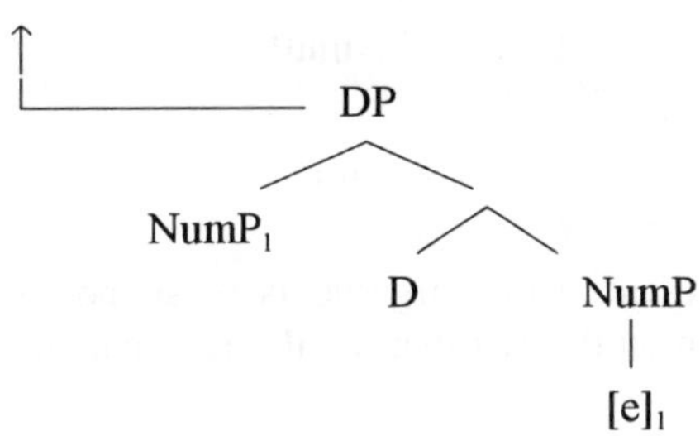

b. *Head movement*

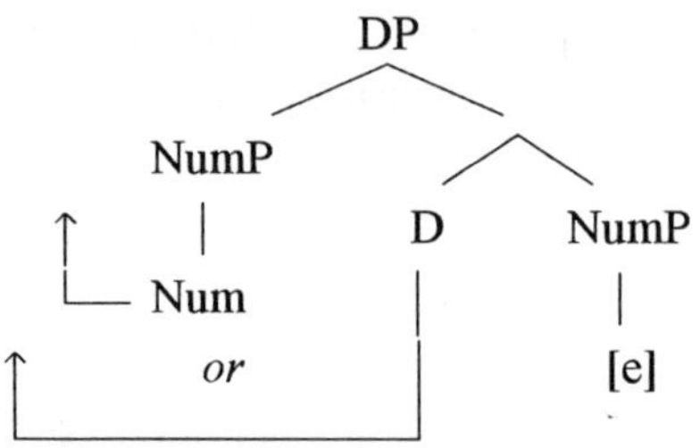

c. *NumP movement and D-standing*:

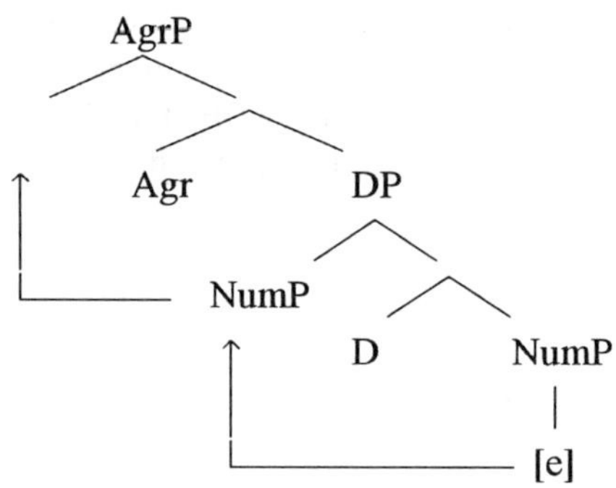

In the remainder of this article, I present strong arguments in support of NumP movement and D-stranding (19c), based on the fact that it will provide a quite simple analysis of Welsh agreement.

4. WELSH AGREEMENT

As in other Celtic languages,[18] there are *agreement* and *non-agreement* configurations in Welsh. Agreement is realized in three different ways depending on the syntactic environment: as an inflectional suffix (subject verb agreement, and prepositional agreement), as a prefix in nominal environments, and as an infix (representing the object between C and V in CVSO structures in highly formal Welsh.[19] Agreement is asymmetric: it is obligatory with pronominals, whether overt or covert, but excluded with lexical DPs. *Pro* is licensed only with overt agreement.

The following examples illustrate the different environments and their particular properties. A summary is presented in (30) below:

(20) a. Agorodd y dynion y drws. (Sadler, p.51)
opened-3S the men the door
'The men opened the door'

b. y dynion a ddarllenodd/*ddralenasant y llyfr. (Sadler, p. 52)
the men pt read-3S / *read-3PI the book
'The men who read the book'

c. Agoront (hwy) y drws. (Sadler, p. 51)
opened-3P (they) the door
'They opened the door'

As these examples show, Welsh finite clauses exhibit VSO order in finite clauses. Finite verbs come in either *analytic* or *synthetic* forms. The analytic form encodes information about the tense and mood of the clause, but is invariantly inflected as third person singular. The synthetic form encodes tense and mood and is fully inflected for person or number. The analytic form must be used with a lexical DP subject (as in (20a)), or with a wh-trace (20b). The *synthetic* form, fully inflected for person and number, *must* occur with pronouns, whether overt or covert. Thus, verbs fully agree with pronominal subjects, but not with lexical DPs. Since third person agreement on Vs does not encode gender difference, it cannot be seen if lexical DPs trigger gender agreement as is the case in Arabic.

Welsh has a class of inflected Ps. A representative agreement pattern for P is given in (21) and (22) for the P *ar* 'on'.

(21) a. ar y plant b. *arno y plant
on the children on-3S the children

(22) Singular Plural (Hendrick, 1988, p.38)

Singular	Plural
1S arnaf (i)	arnon (ni)
2S arnat (ti)	arnoch (chwi)
3SM arno (fe/fo)	arnynt
3SF arni (hi)	

An inflected P does not agree with a lexical DP (21a). Ps differ from finite V which must carry 3rd singular agreement with a lexical DP subject. As (21b) shows, this is impossible with Ps.

Inflected Ps further distinguish themselves from inflected verbs in overtly encoding gender agreement for third person singular pronouns (*arno fe* 'on3SM him' and *arni hi* 'on3SF her'). Gender agreement also shows up in nominal environments *(ei* (3SM) causes lenition).[20]

Agreement shows up as a clitic-like prefix, in the case of genitive agreement in DPs (23), and object agreement with nonfinite verbs (so-called verbal nouns) (24):

(23) a. car Aled
car Aled
'Aled's car'

b. *ei gar Aled
3S car Aled

c. ei gar (hi)
3Sm car Aled

d. *car hi
car her

e. e car (hi)
3SF car (her)

(24) a. Gwnaeth Emrys werthu y llyfr. (Sadler, p.73-74)
did-3S Emrys sell the book
'Emrys sold the book.'

b. *Gwneath Emrys werthu ef.
did-3S Emrys sell it

c. Gwneath Emrys ei werthu (ef).
did-3S Emrys 3SM sell it
'Emrys sold it.'

The agreement pattern is identical to the one discussed so far: agreement is impossible with lexical DPs, but obligatory in pro-drop structures or structures with an overt pronoun. As with Ps, no third person agreement marker appears on N when its subject is a lexical DP.

In highly literary Welsh, an infixed agreement form may *optionally* appear in front of the V when there is a so-called overt presentational suffix (C):

(25) a. Fe'th welodd di. (Sadler (20), p. 75)
pt-2S saw-3S you
'He/she saw you.'

b. Fe'th welodd. (Sadler (21), p.75)
pt-2S saw-3S

The presence of agreement licenses *pro*. If agreement is absent, *pro* is excluded:

(26) Fe welodd *(di).
pt saw-3S you

Pronouns cooccurring with agreement are called "affixed" or "auxiliary" pronouns. These pronouns we have analyzed as agreeing Ds:

(27) Auxiliary pronouns cooccur with agreement.

Pronouns that occur in configurations without agreement, most notably object pronouns in sentences with VSO order, or complements of Ps that lack an inflectional paradigm, are called "independent" pronouns. These pronouns cannot cooccur with overt agreement:

(28) Independent pronouns cannot occur with agreement.

In verb initial VSO structures, object agreement is impossible in all varieties of Welsh: the independent form of the pronoun must be used, and *pro* is disallowed:

(29) a. Glywoch chi Sion/ef.
hear-PAST-2PI you Sion/him
'you heard Sion/him.'

b. *Clywoch chi.
'You heard him.'

c. Glywoch chi mi.
hear-PAST-2PL you me
'you heard me.'

As (29b) shows, *pro* is licensed only if there is overt agreement.

(30) summarizes the distribution of lexical DPs, agreement, and auxiliary pronouns:

(30) (a) overt pronouns and pro obligatorily trigger agreement in agreement environments;
(b) auxiliary pronouns cooccur with agreement;
(c) independent pronouns cannot occur with agreement;
(d) Lexical DPs don't trigger agreement;
(e) *pro* is licensed iff there is overt agreement;
(f) Finite verb forms must carry agreement, either full agreement with a pronoun or third person agreement; other lexical categories have non-agreeing forms;
(g) Finite verb forms do not agree in gender; P and N do.

In this paper, I present an account for the properties in (30a-e). I will not address the interesting problem which concerns the source of third person agreement on tensed verbs in (30f),[21] nor will I attempt to say anything about the problem of the distribution of *a*, wh-traces and the resumptive pronoun strategy (see Sadler (1988), Rouveret (1990) among others).

5. THE ANALYSIS OF WELSH AGREEMENT

Agreement in Welsh is asymmetric: only pronouns trigger agreement, but lexical DPs do not. This suggests that overt agreement reflects the overt syntax, and therefore yields insight into the distribution of different elements in the overt syntax.

(31) Agreement in Welsh reflects the configuration in the overt syntax.

If agreement is triggered in [Spec, Agr] in the overt syntax, as I assume, there must be an agreement triggering element in [Spec, Agr] in the overt syntax with *pro* and pronouns, but not with lexical DPs. Thus:

(32) a. Agr is triggered by an overt pronominal element in [Spec, Agr];
b. There is no agreement triggering element in [Spec, Agr] in the overt syntax with lexical DPs.

This simple account of the agreement asymmetries will receive strong support if it can be shown that lexical DPs do not occur in [Spec, Agr], but pronominal elements do. What pronominal element actually could occur in [Spec, Agr]? There are two possibilities: either the entire pronominal DP moves to [Spec, Agr], as in Koopman (1989), or part of the DP (NumP) occurs in [Spec, Agr], with D stranded in a DP position. I will now motivate the latter.

5.1. Are pronominal DPs in Spec, Agr? Analytical difficulties

Suppose first that pronominal DPs, but not lexical DPs, move to [Spec, Agr] positions in the overt syntax. If this is correct, overt pronouns and lexical DPs should occupy different positions in the overt syntax. Is empirical support for a different distribution of pronominal and lexical DPs?

In this section, I discuss the analytical difficulties that one encounters in trying to argue for pronominal DP movement. These analytical problems are actually quite representative of the kind of problem one runs into doing contemporary linguistic analysis.

It is difficult to determine where a head is in the overt syntax, and where lexical DPs occur. There are many configurations in which pronominal DPs *could* very well occur in higher positions than lexical DPs, but which do not allow one to establish that this *must* be the case. I will briefly discuss these configurations here, and discuss some data in Welsh and other Celtic languages, which suggest that pronominal DPs are higher. Through this discussion, I also illustrate why the argument is difficult to make.

Consider the following configuration, in which both a pronominal DP and a lexical DP raise, but in which the pronoun raises to a Specifier position that is higher than the lexical DP. Suppose furthermore that the head does not end up in the position between the pronoun and the lexical DP, but raises to a position which is still higher than the pronoun[22]:

(33) X [pronoun [e [lexical DP [...

Many configurations in Welsh fit the schema in (33). Consider subjects for example. Lexical subject DPs must raise in Welsh, as argued in Koopman and Sportiche (1991, 234-235), since they precede sentence medial negation. But pronouns could still raise higher to an Agr triggering position. This is only possible if V moved higher than Agr, possibly to C, as is often assumed for Celtic languages.[23] Basically, then, some modifier must be found which occurs between the pronoun and the lexical DP. But even then, the argument is difficult to make, because a different positioning of the pronoun could be due to head movement of the pronoun, and the argument cannot be made tightly.

Awberry (1990) presents evidence based on the distribution of subjects with sentence medial negation in Welsh dialects which suggests that pronouns must always raise. In these dialects, pronouns must precede sentence medial negation, definite DPs may precede or follow negation[24] and indefinite DPs follow negation. However, this paradigm only shows that pronouns must raise; it does not show that pronouns must raise to a yet higher position than DP subjects preceding medial negation.

Other Celtic languages have phenomena that suggest that pronominals occupy different positions than lexical DPs. The problem here is that the distribution might be due to head movement of the pronoun. There is a paradigm in Breton for example which suggests that pronominal DPs move to Agr triggering positions which lexical DPs cannot reach. While Breton disallows overt pronouns with agreement[25], the Welsh pattern surfaces with emphatic pronouns, which obligatorily trigger agreement. The following distribution of DPs, agreement, and emphatic pronouns hold in negative sentences in Breton, which involve a negative C (*ne*) and a sentence medial negation (*ket*)

(34) a. Ne gousk ket ar baotred. (Stump, 1984, (35a))
PCL sleep not the boys
'The boys do not sleep.'

b. Ne gouskont ket. (Stump, 1984, (34a))
PCL sleep3P not
'They do not sleep.'

c. Ne gouskont-int ket. (Stump, 1984, (35b))
PCL sleep3P- they not
'They don't sleep.'

d. *Ne gousk-int ket.
PCL sleep-they not

Lexical DP subjects follow medial negation *ket* (34a), and are thus quite low. Pronouns trigger agreement (34b). Importantly, emphatic pronouns *precede* negation (34c) while obligatorily triggering agreement (34d). This seems quite solid evidence that pronouns raise to a higher position than lexical DPs. The emphatic pronoun could have cliticized onto V, however, and could have moved with V.[26]

Two other potential arguments could be made for Irish, based on the distribution of object pronouns in tensed clauses and the position of overt subject pronouns. Irish pronominal objects tend to occur in clause final position. Chung and McCloskey (1987) analyze this positioning as resulting from a rule of pronoun postposing, which postposes only pronouns to the nonargument maximal projection that most immediately dominates them. There are two surprising facts about this rule: it applies only to *pronominal* DPs, and it shifts a pronoun to the *right*. If pronominal DPs must occur in [Spec, Agr] positions, but Celtic lexical DPs cannot, we can assume that the pronoun postposing is actually an instance of the same process: the pronoun moves to a Spec position which is only available in tensed sentences, but to a *final* Spec position. Although these data are highly suggestive, I do not know why this Spec position would be final, whereas other Spec positions are initial, nor how to motivate its final position through obligatory movement of the complement of AgrO for example. Chung and McCloskey (1987) further note that certain parenthetical or interjective elements can intervene between the finite verb and the lexical subject, but not between the finite verb and a pronominal subject:

(35) a. Tá arndóigh, saighdiũirí ar on bhelach. (C&McC, (133a)
be(pres) of-course soldiers on the road
'There are, of course, soldiers on the road.'

b. *Tá arndóigh, siad ar on bhelach. (C&McC, (134a)
be(pres) of-course they on the road
'They are, of course, on the road.'

They argue that the pronoun is a phonological clitic. However, if pronouns must move to some Specifier position higher than lexical DPs, the pronouns in (35) could actually be in a position higher than parentheticals, which in turn are higher than lexical subject DPs. The problem again is that the pronoun could have cliticized from the surface position of the lexical DP onto the inflected verb, and could have pied-piped with the V to the higher head position. This analysis has some appeal for Irish because the finite verb in these configurations does not agree with the pronoun.

These examples illustrate some of the problems in determining where pronouns occur: either it cannot be ruled out that the head has moved to some head position higher than Agr, or a different distribution of pronouns could be the result of head movement from the position in which regular DPs occur. The question thus arises if it *can* be shown where pronominal DPs occur, and whether our hypothesis is actually testable.

5.2. Pronouns and agreement in DPs: pronominal DPs are not in [Spec, Agr]

It can be shown through the structure of Welsh DPs that pronominal DPs have the same distribution as lexical DPs, and that lexical DPs do not occur in [Spec, Agr] (thus confirming (32b)). This argument requires some preliminary discussion and analysis of the internal structure of the DP in Welsh.

In the DP, Agr surfaces as a prefix which can be separated from the head N by certain adjectives and numerals:

(36) ei tri hen lyfr (Sadler, (48) p. 105)
his three old book
'the three old books'

While prenominal adjectives can be argued to form lexical compounds with N (Williams, 1980, p.38), as McCloskey and Hale (1984, p.524) propose for the prenominal adjectives in Irish, a lexical compound analysis cannot be justified for numerals. Agr is thus an independent head, occurring higher than Num. The head-hood of Agr is further confirmed by the fact that Agr can cliticize onto a preceding N.

The data in (36) might suggest that the head N is lower than Num. The situation is slightly more complex, however, because the head N precedes adjectives, the subject DP and complements:

(37) a. hanes bywiog Wyn am yr ymfudwyr (Awberry (1977), p. 192 (200))
story lively Wyn about the immigrants
'Wyn's lively story about the immigrants'

b. llyfr newyydd Dafydd
book new David
'David's new book'

These examples show that subjects of DPs occur quite low in the structure, presumably in situ in [Spec, NP]. This explains why they precede complements of N. Since they are preceded by N, N must have moved to a position higher than [Spec, NP], at least to Num. If adjectives are adjoined to NumP, N must have moved to a higher head position, possibly Agr. N would therefore right-adjoin to Agr. This is in accordance with the fact that quite generally adjunction to a head in Welsh seems to be *right* adjunction, as can be shown through the syntax of Welsh numerals. Numerals do not appear to behave as modifiers, but rather as heads in Welsh. Plural number is realized as an inflectional suffix on Ns. Welsh numerals, however, must be followed by a *singular* noun (Williams, 1980, p.41); cf. *three book* in (36) not *three books*. Thus *plural nouns and cardinal numerals are in complementary distribution*. This suggests that plural and numerals occupy the same head position (Num). I will assume that N with plural base generated morphology raises to Num, where plural is checked by Num. When Num is occupied by a numeral, the numeral occurs in Num, and there is therefore no way to check plural morphology. The noun therefore cannot carry plural morphology. Since overt numerals (i.e. Num) precedes the head N, N *right-adjoins* to Num[27]. Agr occurs to the *left* of Num, and is therefore higher than Num. I assume that Num, containing Num and N, *right-adjoins* to Agr. Agr possibly excorporates and moves on to D.

(38)

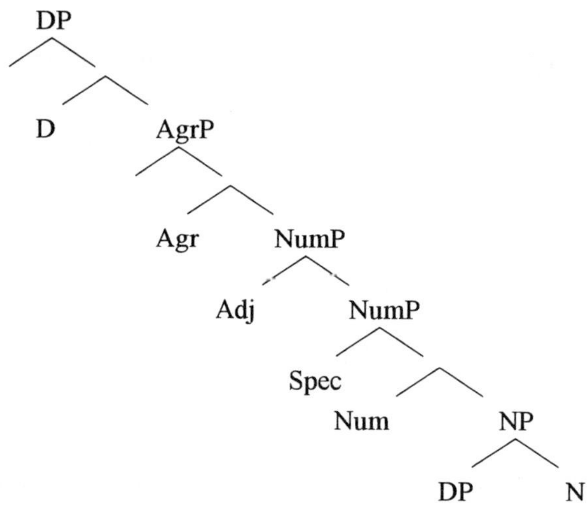

Since the lexical subject DP occurs in [Spec, NP], and not occur in [Spec, Agr], there is no agreement with lexical DPs.[28] Since pronominals obligatorily trigger agreement, there must be a pronominal element in [Spec, Agr] in the overt syntax. We can now test our hypotheses as to what pronominal element raises, DP or NumP. If the entire pronominal DP raises, overt pronouns should precede adjectives; if it does not, pronominal DPs should occupy the same position as lexical DPs.

(39) a. ei hanes (*ef) bywiog *ef* am yr ymfudwyr
Agr3 story (*his) lively his about the immigrants
'His lively story about the immigrants'

b. ei llyfr (*ef) newyydd *ef*
Agr3 book (*his) new his
'His new book'

Overt pronominal elements can never precede adjectives, as the examples in (39) show. They therefore occur in exactly the same position as lexical DPs.[29] Pronominal DPs are therefore not in [Spec, Agr], and we are back to the question why pronominals trigger obligatory agreement, but not lexical DPs.

5.3. NumP movement and D-stranding

Suppose that overt agreement is triggered in [Spec, Agr], not by the entire pronominal DP, but by a NumP, with the overt pronoun (the auxiliary pronoun) being a stranded D. If this is correct, (39b) would be assigned the following representation:

(40)

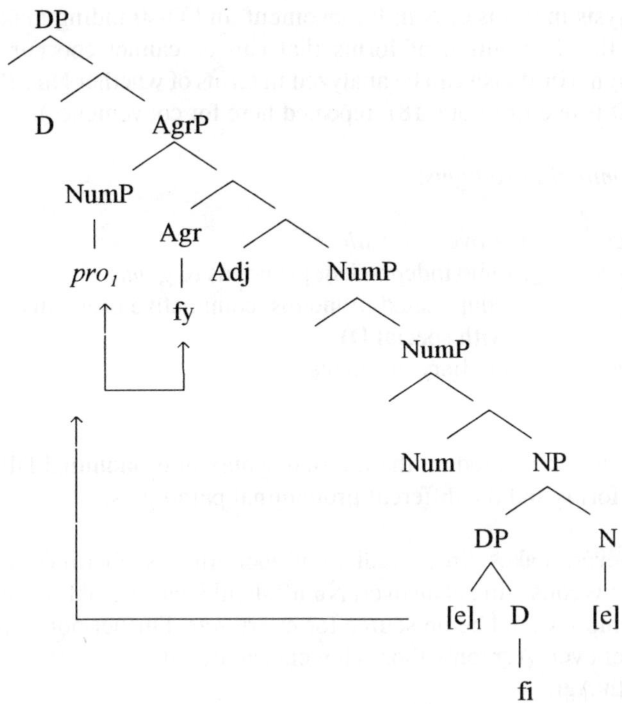

NumP moves DP internally to [Spec, DP], triggering overt agreement on D (*fi*). Then, NumP moves on to [Spec, Agr], triggering agreement on Agr. (It is very likely, but irrelevant for the present discussion, that *fy* further moves on to D, thus accounting for the obligatory absence of the determiner when Agr is present). Note that the form tells us that Agr is overt with NumP silent, rather than NumP overt and Agr silent *f* and *d* characterize first and second person agreement on D, *m* and *t* characterize first and second person NumP. Covert NumP thus occurs in [Spec, Agr], with Agr overt. This actually always seems to be the case: only covert NumP can occur in [Spec, Agr]; there do not appear to be forms that are composed of overt NumP and silent Agr or with both NumP and Agr overt.[30] I conclude tentatively:

(41) Only covert NumP, *pro*, can appear in [Spec, Agr] in the overt syntax.

The situation in (41) therefore slightly differs from that within the DP, where in some cases NumP can be overt. In fact, (41) holds for all Celtic languages: overt Agr cooccurs with *Pro*, i.e. silent NumP.

The agreement asymmetry thus immediately follows from the different internal structure of pronominal and lexical DPs: only with pronominal DPs can NumP move to [Spec, Agr] and trigger agreement. Lexical NumPs cannot escape from the DP simply because N undergoes head movement to Agr: there can therefore be no NumP in [Spec, Agr] with lexical DPs, and lexical DPs therefore do not trigger agreement.

The proposed analysis in terms of NumP movement and D-stranding makes specific predictions about the distribution of forms that can or cannot cooccur with Agr. Pronominal paradigms in Welsh can be analyzed in terms of whether NumP is overt or not, and whether D is overt or not ((18), repeated here for convenience):

(18) *Welsh Pronominal Paradigms*:

NumP	*D*	*+ = overt, - = silent*
+	-	: some independent pronouns (e.g. *mi*)
+	+	: reduplicated pronouns, conjunctive pronouns (with special D)
-	+	: auxiliary pronouns
-	-	: *pro*

This analysis is exclusively based on the internal syntax of pronominal DPs, i.e. on a comparison of the forms and the different pronominal paradigms.

This analysis, however, makes direct predictions about the distribution of NumPs and Ds. Pronominal forms consisting of an overt NumP should *never* be able to cooccur with Agr, since there simply would be no source for overt Agr. Furthermore, since NumP movement underlies overt Agr, only those elements analyzed as stranded Ds should be able to cooccur with Agr.

These predictions are borne out. Pronouns, with overt NumP, like *mi* (1P) for example, can only occur in environments without Agr:

(42) a. gyda *mi*
with me

b. *arnaf mi
on 1PS me

(42b) is ruled out for the simple reason that there is no NumP in [Spec, AgrP]. (42a) is fine since there is no overt Agr.

Reduplicated and conjunctive pronouns contain an overt NumP and should never cooccur with agreement. This again is correct: reduplicated and conjunctive pronouns are defined as independent pronouns, i.e. pronouns that only occur in configurations without agreement.

Overt pronominal elements cooccurring with Agr must necessarily be stranded Ds since NumP has moved away. This is correct; only the so-called auxiliary pronouns, i.e. precisely those pronouns analyzed as agreeing Ds, can cooccur with overt Agr. Some independent pronouns are composed of a null NumP and an overt D. These pronouns can be used as independent pronouns, as in (43a), where the pronoun is used as an answer to a question (the representation of (43a) is given in (43b)):

(43) a. fi 'I/me'
b. $[_{DP}\ [_{NumP}\ \textit{pro}\ [\text{fi}]\ [_{NumP}\ e]]$

Finally, consider the distribution of *pro*, (i.e. silent NumP), which can be summarized as follows:

(44) *DP-internal Pro-drop*

a. $[_{DP}\ [_{NumP}\ \textit{pro}\ [_{AGR}\ \text{Agr}]\ [_{NumP}\ e]]$
b.* $[_{DP}\ [_{NumP}\ \textit{pro}\ [_{AGR}\ e]\ [_{NumP}\ e]]$

DP-external pro-drop

c. $[[_{NumP}\ \textit{pro}_1]\ [_{Agr}\text{Agr}\ [\ldots\ [\ [_{DP}\ [e]_1\ [_{D}\text{D}]$
d. $[[_{NumP}\ \textit{pro}_1]\ [_{Agr}\text{Agr}\ [\ldots\ [\ [_{DP}\ [e]_1\ [_{D}[e]]$
e.* $[[_{NumP}\ \textit{pro}_1]\ [_{Agr}[e]\ [\ldots\ [\ [_{D}\text{P}\ [e]_1\ [_{D}\text{D}]$
f.* $[[_{NumP}\ \textit{pro}_1]\ [_{Agr}[e]\ [\ldots\ [\ [_{DP}[e]_1\ [_{D}[e]]$

With DP internal pro-drop, D must be overt (44a); similarly, with DP external pro-drop Agr must be overt (as (44e) shows it is not sufficient for D to be overt). This paradigm shows that silent NumP must be identified by an overtly agreeing head, or rather that the head of a silent NumP head must be identified by an overtly agreeing head:

(45) The head of a silent NumP chain must be identified by overt agreement.

The analysis is thus extremely simple. Overt agreement is triggered by NumP raising (and when overt agreement is chosen, NumP must raise). Lexical DPs do not trigger agreement, because there is no possible NumP that triggers agreement.

To conclude this section, let me illustrate the derivation. A sentence like (46a) has the corresponding derivation in (46b):

(46) *NumP to [Spec, Agr]; stranded D is overt*

a. Trois fi.
Turn-past-Agr I
'I turned.'

b. $[_{AgrP}\ [_{Num}\ \text{pro}]_1\ [[[_{AGR}\ \text{V}]\ \text{T}]\ \text{Agr}]\ [_{T}\ e]\ [_{DP}\ [_{NumP}\ e]_1\ \text{fi}\ [_{T}\ e]\ [_{VP}\ [_{DP}\ e]\ \text{V}\ldots$

The pronominal DP moves to the position definite DPs must move to in the overt syntax. Within the DP, NumP moves to [Spec, DP] triggering agreement on D, and further escapes to [Spec, AgrP], stranding D. Tensed clauses must have agreement. This forces raising of NumP in the overt syntax. The V itself raises through the relevant intermediate functional categories to T and Agr.

This analysis is compatible with any analysis in which V, or any other lexical category for that matter, never raising higher than Agr. This is consistent with McCloskey's (1992) arguments that C in Irish lowers to V (i.e. V is not in C). As far as I have been able to determine, similar arguments hold for Welsh.

5.4. Conclusion

The analysis of the internal structure of Welsh pronouns and the external distribution of pronouns, agreement, *pro* and lexical DPs provides strong evidence both for the proposal that pronominal DPs consists of a NumP in [Spec, DP] and a D which are in an agreement relation, and for the existence of NumP movement to a higher [Spec, Agr] position where external agreement is triggered. It is precisely the richness of pronominal paradigms, and the occurrence of asymmetric overt agreement, which allowed me to construct an argument for NumP movement resulting in overt agreement. This analysis furthermore presents strong evidence in favour of agreement as a Spec head relation.

Let me consider more concretely how the generalizations in (30), repeated here for convenience, are accounted for.

(30) (a) overt pronouns and *pro* obligatorily trigger agreement in agreement environments;
(b) auxiliary pronouns cooccur with agreement;
(c) independent pronouns cannot occur with agreement;
(d) Lexical DPs don't trigger agreement;
(e) *pro* is licensed iff there is overt agreement.

Obligatory (silent) NumP movement explains (30a). With overt agreement, NumP *must* move to [Spec, Agr] in the overt syntax.

Auxiliary pronouns cooccur with agreement (30b), (i.e. there is clitic doubling with overt pronouns), because they are agreeing Ds, stranded by NumP movement to [Spec, Agr].

The statement in (30c) has to be adjusted somewhat: those independent pronouns that contain an overt NumP cannot cooccur with agreement. This explains why reduplicated pronouns, conjunctive pronouns, and simple independent pronouns with overt NumP do not cooccur with agreement: NumP is overt within the pronominal DP in all these pronouns, and (sentential) Agr can only cooccur with silent NumP (41).

Lexical DPs do not trigger agreement, (30d), simply because they do not occur in [Spec, Agr] in the overt syntax. Their NumP is within the lexical DP (the head N moves to Num, Num and N move to Agr), and therefore there is no possible source for agreement. Agreement asymmetries therefore follow entirely from the structure of the DP in the overt syntax: if NumP is in [Spec, DP], further movement is possible; if NumP is not in [Spec, DP], as in the case of lexical DPs, further movement is of course impossible.

Pro (empty NumP) is licensed iff there is overt agreement (30e). The head of an empty NumP chain must be licensed by overt agreement. *Pro* thus occurs in [Spec, DP] licensed by an agreeing D, (this is the case with certain independent pronouns that consist of an agreeing D in contexts without overt agreement), or when licensed by overt agreement. Note furthermore that we have further assumed that the lexical head occurs in Agr in the overt syntax: heads move to Agr, but no higher. Head movement will extend the domain, allowing NumP to reach [Spec, Agr]. Since heads do not move any higher, however, NumP cannot escape outside of these domains: there are therefore no clitic climbing or cases of multiple agreement beyond those already discussed.

The proposed analysis strongly supports agreement as a reflex of an overt Spec head relation. There is a competing theory for agreement, however, which consists of treating agreement as an incorporated pronoun (Anderson (1981), McCloskey and Hale (1984), Taraldsen (1992), Rouveret (1991)). This theory is motivated primarily on the basis of Celtic languages, and aims to capture the complementary distribution of agreement (no full agreement with lexical DPs, full agreement with pronominal DPs). I have offered a different account for this asymmetry which is based on a fundamental asymmetry between the overt distribution of pronouns and lexical DPs in phrases (both clauses and DPs). This asymmetry is in fact widely attested in other languages, and the present theory is therefore simpler in so far as it makes use only of Spec head agreement and the distribution of pronouns. Note that I cannot actually show that the incorporation theory of agreement is incorrect: this theory would be consistent with the analysis in this paper, provided that NumP movement to a position from which its head can be incorporated into some higher head position is assumed (i.e. agreement in DPs is much to high for direct incorporation). Proponents of the incorporation theory could capitalize on the fact that NumP triggering agreement must obligatorily be empty: if the head of NumP incorporated, this would follow. However, if this is correct, there would be several ways of triggering agreement in Universal Grammar, surely not an attractive hypothesis from the point of view of the language learner.[31]

5.5. NumP movement in Irish and other Celtic languages

The NumP movement analysis quite straightforwardly extends to Irish and other Celtic languages: NumP movement must occur in configurations with overt agreement, but cannot occur in the absence of agreement.[32] Thus, full pronouns must be used in non-Agreement contexts.

As McCloskey and Hale (1984) discuss, Irish verbal paradigms are defective and do not show full agreement for all forms. When an analytic verb form exists, or is chosen, full pronouns must be used, i.e. there can be no NumP movement, because silent NumP is not identified by an agreeing head. Although Irish differs from Welsh and lacks pronominal forms consisting of a NumP and an agreeing D, there are other forms which consist of a pronoun, i.e. NumP, preceding a particular head (reflexive or emphatic pronouns, demonstratives,[33] and contrastively stressed pronouns (McCloskey and Hale (1984), pp. 493-494.). Only the contrastive suffix overtly agrees with NumP in its Specifier. These heads can be stranded by NumP movement as well, in configurations with overt agreement (cf. McCloskey and Hale (1984)). Note that in non-agreement contexts, the NumPs *precede* these heads, as expected: since NumP is in the Specifier of the particular head, it has precisely reached the position from where it could undergo further movement. As in Welsh, the stranded head occupies the same position as lexical DPs,[34] as the internal structure of DPs allow us to conclude:

(47) a. a shaol suarach féin. (M&H, 1984, (63b),p. 514)
MRS life wretched REFLEX
'His own wretched life.'

b. ár soal stoirneach-inne. (M&H, 1984, (65c),p. 515)
P1 life stormy CONTR P1
'Our stormy life.'

Quite generally, lexical DPs cannot trigger agreement, for the same reasons as in Welsh: lexical NumPs do not move to [Spec, DP] in the syntax. They therefore cannot move to [Spec, Agr]. There is no possible source for overt Agr.

6. MAINLAND SCANDINAVIAN NumP MOVEMENT

So far, Celtic languages illustrate silent NumP movement to [Spec, Agr], with overt Agr. The NumP movement analysis also straightforwardly accounts for the distribution of pronouns in Mainland Scandinavian languages, as I will now show. As is well-known (Holmberg (1986), among others), pronouns in the mainland Scandinavian languages move to positions where lexical DPs cannot occur. Pronoun shift depends on V-movement, in the sense that the pronoun never moves to a position outside the domain of a verbal chain. This is generally taken to indicate that V-movement extends the domain in which movement is possible. Because pronouns fail to license parasitic gaps, pronoun shift is generally assumed to be movement to an A-position.

I will analyze Norwegian pronouns in this section, although I believe that the results extend to other Mainland Scandinavian languages, as well. Let us start with an analysis of Norwegian pronouns, based on Hestvik (1990). Hestvik argues that English pronouns and Norwegian pronouns, which are neither homophonous with Ds nor transparently composed elements, have different representations: English pronouns are undecomposable NPs (DPs in more modern terminology), and cannot be modified (**he* *with long hair, *she from upstairs*), Norwegian pronouns are Ns and behave as Ns in the sense that they can be modified. The equivalent of *he with long hair* is fine in Norwegian. Translating Hestvik into our structures for pronouns, pronouns are either N or NumP, depending on whether the pronoun can be modified by a PP or not. Treating English pronouns as undecomposable NumPs and Norwegian pronouns as decomposable into N and Num, is rather unsatisfactory, however. After all, the NumP projection expresses the idea that there is both a Num head and a N head. English pronouns should therefore be treated as being composed of N and Num (and person) as well. The question thus arises how to account for the possibility/impossibility for a pronoun to be modified. Let us maintain the account that English pronouns start out as N as well, and explore a different way to account for the possibility of pronominal modification. Assume that it follows from the particular overt syntax of the DP. Within the proposed pronominal structure, there are two likely candidates: PP modification is impossible if N movement occurs, or, PP modification is impossible when NumP movement occurs. The former will not explain the impossibility for modification. First, in comparable configurations V movement seems perfectly possible. And secondly, in languages with head movement of N, like French for example, N can move and still be modified by a PP modifier.

(48) L' ami italien de Jean aux cheveux longs
the friend Italian of John with hair long
'John's Italian friend with long hair'

The head N has raised out of the NP to a position higher than the position in which the possessor occurs, and the adjective. Yet PP modifiers can occur.

If the presence of a PP modifier does not depend on N movement, could it depend on overt movement of NumP? For this option to be right, it must be shown that PP modifiers may occur when NumP does not move in the overt syntax, as is the case with regular English DPs, and that PP modifiers are excluded if NumP moves to [Spec,DP], as with English pronouns.

The possibility of NP modification by a PP is not restricted to pronouns cross-linguistically, but actually also shows up with lexical Ns. While Ns in most Indo-European languages can be modified by PPs, Ns in many other languages cannot (Abe (Kwa) (Koopman, field notes), Bambara (Mande) (Crook (1990)), Nweh (Nkmenji (1993)), and Turkish (Kural, personal communication)). In these languages, PP modifiers occur with verbal projections, but not with nominal projections[35] (the equivalent of PP modification is expressed by a relative clause). Why then are PP modifiers allowed in certain languages, but not in others? Languages that do not tolerate PP modifiers of NP appear to be D final. Often D agrees with the head N. Since many of these languages are otherwise head initial, this suggests that a projection of N has moved to [Spec, DP], triggering agreement on D, as Nkemnji (1995) argues for Nweh (Grassfield Bantu). If this is correct, the structure is roughly identical to the one proposed for English pronouns, but with lexical NumP movement:

(49)

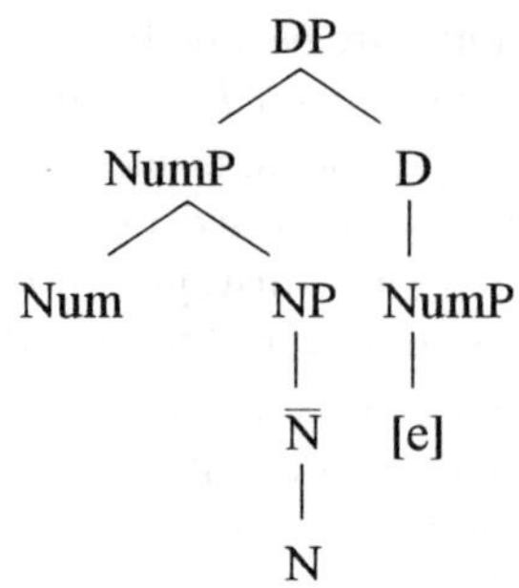

Given these observations, we can now assume the descriptive generalizations in (50):

(50) a. PP modifiers are not allowed if NumP moves to [Spec, DP].
b. PP modifiers are allowed if N moves to Num or a higher head, or if N is in-situ.

Although it is unclear at this point what explains (50a), I conclude that PP modifiers cannot occur with English pronouns, not because they are unanalyzable NumPs, but because NumP obligatorily moves to [Spec, DP]. The impossibility of PP modification thus provides an additional argument for NumP movement in English pronominal DPs.

Since Norwegian pronouns can be modified, NumP has not moved to [Spec, DP] when the pronoun is modified. This raises the question if NumP ever moves in Norwegian. I assume it does, because of the fact that NumP can undergo movement to a higher Spec position, yielding pronoun shift: this indicates DP internal NumP movement. Norwegian pronouns are thus analyzed as consisting of a silent D and an overt NumP, with optional DP internal movement of NumP to [Spec, DP]. However, when the pronoun is modified, it must be the case that overt NumP movement to [Spec, DP] has not taken place. Since NumP is overt, overt NumP should be able to further move to [Spec, Agr] positions, as in Welsh, but now with Agr silent: this, I assume is precisely what happens with pronoun shift. NumP movement should furthermore only be possible when NumP moves to [Spec, DP] in the overt syntax. In particular, if NumP does not or cannot move to [Spec, DP], the pronominal DPs should have exactly the same distribution as lexical DPs.

These predictions are borne out. Although pronoun shift is strongly preferred, it is not obligatory (it is near obligatory). This immediately follows from the optionality of NumP to [Spec, DP] movement in the overt syntax. If NumP does not move to [Spec, DP], a pronominal DP will have the same distribution as a lexical DP, because NumP will be unable to move to [Spec, Agr] in the overt syntax. This yields the effect of optionality of pronoun shift, which thus in fact reduces to the distribution of constituents within the DP. If NumP *does* move to [Spec, DP], it will be able to undergo further movement and move to [Spec, Agr], thus yielding object shift.[36] Pronoun shift should be *impossible* in those cases where NumP movement to [Spec, DP] is impossible, i.e. with pronouns that are modified by PPs. Holmberg (1986) notes (but does not explain) the fact that pronouns that are modified by PPs cannot undergo object shift. The following paradigm shows this for Norwegian[37, 38]

(51) a. Jeg liker *ham* ikke.
I like him not

b. Jeg liker ikke *han med langt haring.*
I like not him with long hair
'I don't like him with long hair.'[39]

c. *Jeg like *han med langt haring* ikke.
I like him with long hair not

d. *Jeg liker *han* ikke *med langt haring.*
I like him not with long hair

(51a) involves DP internal NumP movement to [Spec, DP] followed by NumP movement to [Spec, Agr]. (51b) shows that a pronoun with a PP modifier has the distribution of a lexical DP: it cannot undergo pronoun shift (51c), nor, quite surprisingly, can the pronoun shift stranding the modified PP (51d). This paradigm is immediately explained: when a NumP is modified it is prohibited from moving to [Spec, DP] (50); (51c) is therefore excluded in the same way as shifting of a lexical DP, i.e. NumP cannot reach [Spec, AgrO]; (51d) is excluded in the same way: NumP is unable to reach [Spec, DP] in the first place, and movement to [Spec, Agr] is therefore impossible.[40]

A final property of object shift is that it fails to license parasitic gaps. This could in fact be independent of the A or A-bar status of the landing site, but due rather to the fact that the moved category is too small, i.e. NumP, not DP, to license a parasitic gap.

7. ENGLISH PRONOUNS

Although the distribution of lexical DPs and pronouns in English overlaps to a large extent, their distribution differs in a number of constructions. As I will now show, the distribution of English pronouns also reduces to the movement of NumP to some designated position (which I assume is [Spec, Agr], as in Welsh and in Norwegian). Consider first, however, what to expect given the particular structure of the English pronominal DP. English pronouns are composed of overt NumP and silent D (7). If there is further NumP to [Spec, Agr] movement, as in Welsh and Norwegian, *overt* NumP should occur in [Spec, Agr]. Moreover, given the fact that NumP is *overt*, (and not silent as in Welsh), we expect Agr to be *silent*, given the fact that English seems to mark either Spec or head position, but not both (i.e. the doubly filled Comp filter):

(52) a. $[_{\text{NumP}}$ pronoun $[_{\text{Agr}}$ e] ... (*English, Norwegian...*)
b. $[_{\text{NumP}}$ [e] $[_{\text{Agr}}$ Agr] ... (*Welsh, Irish ...*)

If this is correct, English is not a pro-drop language, simply because NumP is overt not silent. This does not explain why NumP cannot also be *covert* in configurations that *do* show agreement. Suppose however that *pro* (i.e. covert NumP) is only licensed if the head of NumP is in a Spec head relation with overt Agr (or rather with the head of an overt Agr chain), as in Welsh. This configuration is arguably never satisfied in English (except for the [Spec, CP], and C of relative clauses, where NumP can actually be empty): English has no person agreement other than some restricted subject verb agreement. Agreeing main verbs are always lower in the structure than AgrS; AgrS is therefore always empty at S-structure, and NumP must therefore be overt. If auxiliaries carrying Agr never occur in AgrS either, but either remain lower, or raise higher (in the case of Subject-Aux Inversion),overt Agr is actually never in a position to license the head of an empty NumP chain. NumP therefore must always be overt.

It follows from (52b) and the discussion above that there will be no analytical help in English from the presence or absence of agreement; nor can we use stranded Ds, because D is silent. Arguments for NumP movement can be constructed, though, on the basis of the syntax of certain QPs containing *[all, both]* (henceforth *allPs*), which can be stranded. Consider the following paradigm:

(53) a. All of these people/all of them came.
b. All these people/*all they came.
c. I saw them all/*these people all.
d. I spoke to them all/*to these people all

All can be followed by an *of-DP* (53a) (which has no bearing on our discussion), or by a bare DP (53b), which is the configuration that interests us. When *all* is followed by an *of-DP*, pronouns and lexical DPs distribute in exactly the same way as they do in PPs. When *all* is followed by a bare DP, as in (53b), pronouns cannot follow *all* but must precede (53c). The distribution of pronouns and lexical DPs is thus asymmetric: pronouns must precede *all*, but lexical DPs cannot (53c, 53d). Based on this paradigm Maling (1976) proposes that pronouns and certain quantifiers obligatorily undergo a rule of Quantifier Pronoun flip (*Q-flip*). How should Q-flip be analyzed? What position does the pronoun occupy? There are two possibilities: the pronoun could be either within the projection containing *all*, call it *allP*, as (53d) seems to suggests, or it could be outside the *allP*.

Suppose first that pronouns are within the *allP*, and occupy [Spec, *allP*]. The problem arises why this position is restricted to pronouns: [Spec, *allP*] does not really seem to be restricted to pronouns, in the sense that DPs can move through this projection in the case of Q-float Sportiche (1988). In Celtic languages, lexical DPs can never move through [Spec, Agr] positions, as can be concluded from the fact that full agreement is impossible in case of wh-extraction. Suppose however that, notwithstanding the appearances, the pronoun is not within the *allP*, but rather in [Spec, Agr]. More precisely, suppose that the situation is exactly as in Welsh, with NumP moving to [Spec, Agr]. This must be the analysis, as I will argue in the remainder of this section.

First, support for the fact that the pronoun is outside the *allP* comes from the following data:

(54) a. All these candies I really like.
b. *Them all I really like.

(55) Q. Which ones do you want, all the yellow candies, or all the green ones?
A: a. all the candies
b. all of them
c. *them all
d. I want them all

(56) Q. Which book do you want? This one or that one?
A: a. Both the books
b. Both of them
c. *them both
d. I want them both

While the order *all DP* can be used as an independent DP constituent, the order *pronoun Q* cannot.[41] This constitutes strong evidence that pronouns are not within the *allP*, but actually have raised outside. (57) thus follows:

(57) Pronouns must raise outside the *allP*.

The problem now arises how to reconcile (57) with (53d). I will assume that the pronoun in (53d) has raised to [Spec, AgrP], with the lexical DP in a lower position, and P raising to some projection higher than the pronoun. (This would be possible, if one assumes PP shells, as in Koopman (1991)).

If pronouns are not within the *allP*, where are they? In which position do pronouns occur? Insight into this problem comes from the environments that allow for pronouns linked to *allP*s: subjects, objects or objects of P. Given the analysis of Welsh, we can therefore assume that the pronouns move to [Spec, AgrS], [Spec, AgrO] and [Spec, AgrP] in the overt syntax. But why do only pronouns raise to [Spec, Agr] positions in the overt syntax? The same answer can be given to English as we provided for Welsh and Norwegian. The asymmetry reduces to an asymmetry in the DP structure of pronouns and lexical DPs. Only pronouns move because NumP is in [Spec, DP] in the overt syntax, and can therefore undergo further movement in configurations where [Spec, Agr] is structurally available. Lexical NumP are *in situ* and can therefore not undergo any movement.[42]

We thus assume that, as in Welsh, NumP must move to [Spec, Agr], whenever possible:

(58)

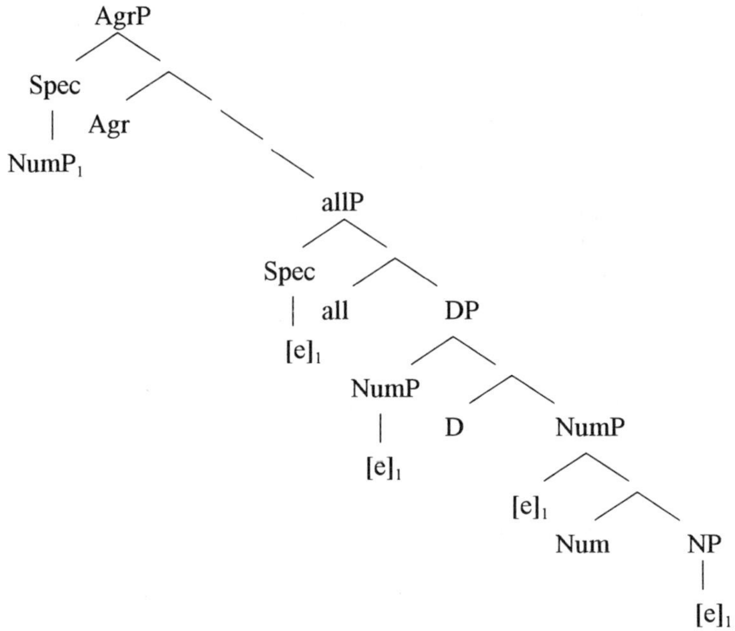

This explains the asymmetry: pronouns are NumPs which appear in [Spec, DP] in the overt syntax. They therefore are in a position from which further syntactic movement not only is possible, but actually required in configurations in which Agr is projected.[43]

The syntax of *allP*s thus shows that pronouns (i.e NumPs) must occur in [Spec, Agr] positions which cannot host lexical DPs (i.e. *with these people all*). If this is correct, it can be concluded that pronouns *always* occur in [Spec, Agr] positions. Support for this assumption comes from the verb particle construction: in the verb particle construction pronouns must occur in pre-particle position, whereas lexical NPs can either precede or follow:

(59) a. I put him up.
b. *I put up him.
c. I put my family up.
d. I put up my family.

If NumPs must occur in [Spec, Agr,] and lexical DPs cannot, there must be both a [Spec, Agr] position and a lexical DP position preceding the particle (as concluded in Johnson (1990)). NumPs must move to [Spec, Agr], and lexical DPs may optionally move to a Spec position preceding the particle. How does NumP movement proceed? The following examples yield insight in the derivation:

(60) a. I called them all up.
b. *I called these people all up.
c. *I called them up all.

The *allP* obligatorily precedes the particle when NumP extraction has occurred. This position is not explained by head movement of *all* left adjoining to P: if P adjoined to the left of the *allP*, the *allP* should be transparent for DP movement, and lexical DP movement to the pre-particle position should be possible. (60b) is therefore wrongly predicted to be possible. I therefore conclude that NumP extraction is preceded by DP movement to preparticle position:

(61)

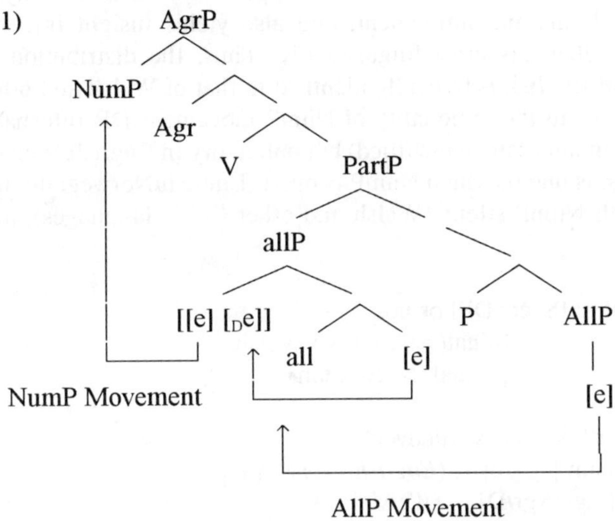

Finally, since English V in the verb particle construction obligatorily precedes pronouns it follows that V moves from a V position to some higher head position, which is higher than the [Spec, Agr] position to which the pronoun moves.

In sum, English pronouns occur in a position from which lexical DPs are excluded. Since distributional asymmetries show up exactly in the configurations where pronouns in Welsh trigger agreement, we argued that English NumPs overtly move to [Spec, Agr] positions when these positions are syntactically projected. Note that we have not entertained an analysis which accounts for the position of pronouns by head movement of the pronoun (NumP movement to [Spec, *allP*] followed by adjunction to the right of a higher host (V, P, and AgrS?).[44] English pronouns do not look like clitics.

A strong argument in favour of the NumP analysis is that it captures the similarities between English and Welsh: English exactly looks like Welsh, except that it has overt NumP, and silent Ds and AgrS. English and Welsh syntactic configurations are thus entirely identical, and the difference ultimately reduces to the internal structure of the pronominal DP: in English NumP is always overt and D and Agr are silent, NumP in Welsh can be empty and D and Agr are overt.

8. CONCLUSIONS

In this article, I have provided arguments that pronominal DPs consist of NumP in [Spec, DP] and D, and shown that this hypothesis has direct consequences for the syntactic distribution of pronouns. In particular, I have argued that asymmetries in the distribution of pronouns and lexical DPs follow from NumP movement to [Spec, Agr] positions, whenever possible. This analysis finds strong support in the internal analysis of Welsh pronominal paradigms and agreement, and also yields insight into the distribution of pronominal elements cross-linguistically. Thus, the distribution of pronouns in Norwegian and English is basically identical to that of Welsh and other Celtic languages, and reduces to the optionality of NumP movement DP internally (optional in Norwegian (pronouns can be modified) but obligatory in English/Welsh), and to whether the language is one in which NumP is overt (English/Norwegian) and D/Agr silent) or is one with NumP silent (Welsh and other Celtic languages), and D/Agr overt:

(62) a. NumP movement to [Spec, DP] or not:
$[_{NumP}$ $[_{D}$]...] obligatory: English/Welsh
optional: Norwegian

b. NumP overt/covert; Agr/D overt/covert:
(i) $[_{NumP}$ pronoun $[_{Agr/D}$ e] ... (*English, Norwegian...*
(ii) $[_{NumP}$ pro [Agr Agr/D] ... (*Welsh*[45] ...)

The coexistence of both systems, as in the Welsh pronominal DP, in which all possible combinations of overt/covert NumP and overt/covert D are attested, provides the empirical basis on which NumP movement to [Spec, Agr] could be established. If there are languages in which NumPs never raise to [Spec, DP], we would expect pronouns to have exactly the same distribution in the overt syntax as lexical DPs, and to not interact in any way with [Spec, Agr], simply because they are too low in the structure.

Our analysis provides new understanding in the syntax of pronouns and agreement, and shows that a common structure underlies both. The next question is how this proposal applies to clitic constructions. Romance clitic constructions quite straightforwardly fit into the structure presented so far. Sportiche (1995) argues for an analysis of clitic constructions as involving a base generated head (the clitic) and an element in the [Spec, Clitic] which has undergone movement from a DP position: if the moved element is actually NumP, Romance languages fit into the schema above, with NumP silent and Agr overt (52bii), although quite interestingly, D in the pronominal DP is silent (or alternatively, has undergone movement to the Agr position). If this is correct, there is thus a common structure underlying the distribution of pronouns, clitic and agreement, involving NumP movement and Agr.

The analysis also raises many new questions which go beyond the scope of this article. I will briefly address some of these. First, the analysis yields new insight in the distribution of *pro*. As a prerequisite for a pro-drop language, the language must be a silent NumP language. Furthermore, the language must be an overt Agr language.[46] And finally, we speculated, the head of Agr must be in a Spec head relation with its Specifier, i.e. pro-drop is dependent on the position of the head carrying Agr.[47]

Agr licenses NumP in the syntax. This raises questions about its nature, its function, and its position (see Sportiche (1995) for discussion of these issues). Agr, as used in this paper, encodes person features, and is identical to what is sometimes called pronominal agreement. The content of Agr differs from participle agreement or adjectival agreement in Indo-European languages, which do not show overt person agreement. Note that I have also been using agreement as a reflex of a Spec head relation, as for instance the DP internal agreement between NumP and D. This suggests that the two views in the literature on Agr are both correct: there are certain instances of Agr that involve an overt syntactic head (i.e. pronominal agreement), and certain cases in which agreement is a byproduct of a particular Spec head relation.

In the languages under discussion, [Spec, Agr] can only host NumP in the overt syntax, but no lexical DPs. This raises two questions. Is this always the case? And if so, why? I will assume without further evidence that the agreement asymmetries reflect LF; Agr only licenses NumP, never DPs:[48] this would follow if the LF function of Agr is the licensing of *pronominal* features. If this is correct, lexical DPs with no pronominal features to license occur in a different position at LF, probably [Spec, NomP,] [Spec, AccP], and contrary to common assumptions, Agr and Case are two independent heads, having distinct licensing purposes (Sportiche (1995)). This would provide a motivation both for the obligatory raising of subject lexical DPs in Welsh and for additional NumP movement: raising of the lexical DPs to the [Spec, Nom] position, and NumP to [Spec, Agr] in the overt syntax, reflects the actual licensing positions at LF. A similar explanation extends to English objects: they can optionally move to [Spec, AccP], but NumP must obligatorily move to [Spec, Agr]. Why does [Spec, Agr] only host NumP, not DP? We will leave this question open for future research.

Where precisely is Agr located and what cross-linguistic differences are observed? Given the position of Agr in Welsh, we can conclude that AgrS occurs just below C, AgrN just below D, AgrP just below the "complementizer" position of P. The position of (pronominal) AgrO is less clear. It is lower than AgrS. Sometimes it seems to be higher than T. as for example in languages in which tense is realized independent of AgrS and AgrO, and in which AgrS AgrO are fused as portmanteau morphemes; sometimes it is lower than T (Basque, Bantu), and separated from AgrS by T. The problem of the position of AgrO thus recalls similar problems concerning the position of NegP.[49]

NOTES

* This paper is dedicated to the memory of Osvaldo Jaeggli. I have been interested for a long time in the distribution of pronouns, and the relation between pronouns, clitics, and agreement (Koopman (1984, 73-76), Koopman and Sportiche (1982) and Koopman and Sportiche (1989)). This paper grew out of a presentation given at MIT in 1987 ("Restrictions on Spec positions"), where I proposed that pronouns but not lexical DPs occur in particular Spec positions in the overt syntax, and ideas about the internal DP structure of pronominal DP (Koopman, 1991). Special thanks to participants of several seminars at UCLA (Spring 1989, Spring 1991, and winter 1993) where part of this material was developed, to audiences at UCLA and Geneva, to Randall Hendrick, James McCloskey, Richard Sproat, for e-mail conversations on Welsh, to Ian Roberts for getting me help with judgments on some crucial Welsh examples, to Arild Hestvik for discussions on Norwegian, and to Kyle Johnson, Tim Stowell and Dominique Sportiche for comments on a previous version of this article (June 1993). The present article is to a large extent identical to the December 1993 version.

[1] This seems to have changed in the beginning of the nineties. See several abstracts in the GLOW newsletter (1993), Rouveret (1990), Cardinaletti (1993), Carstens (1993) among others.

[2] Or some D-like head. The label is not important, the position within the hierarchy of the DP is.

[3] For simplicity reasons, I omit person and gender features, though I believe these should project a person and gender projection respectively. These projections probably enter into an account for the difference in distribution between 1st/2nd personal pronouns and third person pronouns, but do not affect the main points made in this paper.

[4] For example, N to Num movement bleeds NP to [Spec, DP] movement: the NP would contain an unbound trace in the head position. In other words, NumP movement to [Spec, DP] is forced in this case.

[5] This generalization does not extend to emphatic pronouns.

[6] Possibly through an intermediate position if this is the position where person features are licensed (PersP or AgrP). Note that it might be the case that this node is only present when non-third person features must be licensed. This might give some insight into the as yet unsolved problem why in so many languages the distribution of first and second pronouns differs from third person pronouns.

[7] See also Moritz (1993) for Ncufie (Grassfield Bantu).

[8] I will not attempt to find an explanation for this asymmetry.

[9] This may also explain why only NP can be a complement in morphology, but not NumP: only when a morpheme takes a NP complement can N incorporate; the presence of any higher projection will block incorporation. This approach quite generally extends to derive incorporation asymmetries from the structural properties of the overt syntax. (see Koopman (1994) for some discussion), and derives the Relativized Minimality effects for incorporation of Baker and Hale (1990).

[10] I believe that quite generally reduplicated pronouns, which frequently occur in African languages, fit into this structure. It is interesting to note that superficially speaking African languages allow both Spec and head position to be overt at the same time. This is generally not the case in Indo-European languages.

[11] The analysis of Welsh pronominal paradigms is based primarily on Williams (1980).

[12] Williams (1980, p.46) writes: "*the reduplicated forms are somewhat more emphatic than the simple*" I interpret this statement as evidence that there is no real difference between simple and reduplicated independent pronouns

[13] I assume, without arguments, that vowel reduction accounts for the appearance of *y* and not *i* in the paradigm of the reduplicated pronouns.

[14] Although *efe* might be composed of *e* and *be,* I have not been able to come up with anything interesting for third person (non-feminine) pronouns. Arguments will be based primarily on the distribution of first and second person pronouns.

[15] See Morris-Jones (1931), "*The affixed pronoun has often an accent of its own. In poetry it is usually accented.*" (op. cited, p. 83).

[16] A number of irregularities are glossed over: third person pronouns (non-feminine) do not appear to allow for an overt NumP in [Spec, DP]. Some other forms in the auxiliary pronouns do not occur as reduplicated pronouns (*i, ti, hwynt*). The use of both *i* and *ti* appears to be phonologically conditioned (Williams, p.50, 74 (i)). The occurrence of *hwynt* as an auxiliary pronoun is potentially problematic, but not attested in any of the other grammars I have consulted. I will ignore this form in the remainder of this paper.

[17] Conjunctive pronouns have a variety of uses: they are used in apposition to a noun (*David 'he too'*), in case of a conjunction of a lexical DP and a pronoun, and have a variety of other uses (*even I*, *while she*, etc.).

[18] For the distribution of agreement, *pro*, and pronouns in Celtic languages, see, among others, Anderson (1982), McCloskey and Hale (1984), Stump (1984), (1989), Hendrick (1989), Sadler (1988), Rouveret (1990), among others.

[19] This context differs from the others: agreement is only possible in highly formal varieties of Welsh; moreover, it is optional in these varieties. There appear to exist further structural restrictions on when it may appear (Sadler, 1988, p.76).

[20] I have nothing to say about this asymmetry with respect to gender agreement.

[21] I believe that this asymmetry is ultimately due to tense: Tense causes there to be a pronoun in [Spec, TP], which will raise to [Spec, Agr], triggering third person agreement. There is in fact indirect evidence for a null agreement in DPs containing a genitive DP. In this construction, the definite article must be silent, even though the genitive DP is quite low in the structure (in [Spec, NP] see 5.2.). The absence of the definite article can be explained if there is a zero Agr in this construction that must move and substitute for D.

[22] Similar difficulties arise of course when the head stays lower than lexical DPs and pronouns, as is the case with English subjects.

[23] This analysis is not possible if C lowers to I in Welsh, as McCloskey (1993) convincingly shows for Irish. The analysis arrived at below is consistent with V being in Agr, as in Koopman and Sportiche (1991).

[24] I do not know if the position of the DP correlates with a specific reading for the DP or not.

[25] See Anderson (1982), Stump (1984) and Hendrick (1988), among others.

[26] If emphatic pronouns obligatorily moved to [Spec, AgrP], (i) should be impossible, with synthetic verb form:

(i) * ne gouskont ket int
PCL sleep-3P not they

Stump (1989, p.414) points out that this surface string is acceptable, but only with *int* as a right-dislocated element. Borsley and Stephens (1989, p.414) argue however that such examples are grammatical in Stephens dialect:

(ii) ne oant ket int deut. (Borsley and Stephens (1989, p.414))
PCT were3P not they come

See Stump (1989, p. 437 footnote 7) for further discussion. If (ii) is fine in Stephens dialect, it suggest cliticization in Stephens dialect is optional. The grammaticality of (ii) poses no problem for the analysis presented below. It simply implies there are two processes: NumP movement, and D cliticization, with the latter optional in Stephens dialect.

[27] This is further confirmed by the position of the head N in composite numbers:

(i) tair merch ac ddeg
three(fem.) girl and ten
'thirteen girls'

The head N adjoins to the right of the head of the compound numeral.

(ii)

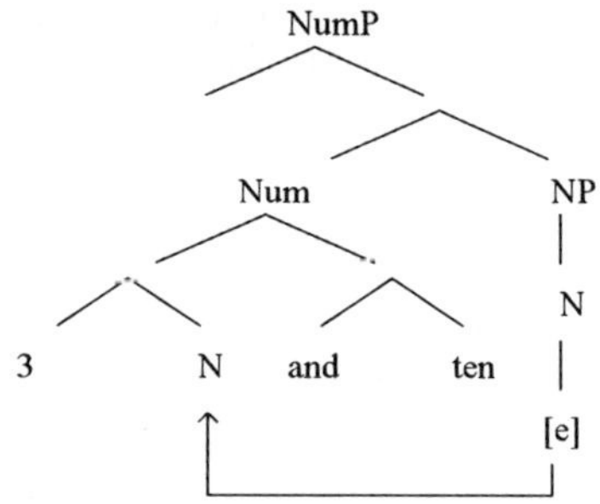

[28] Agr does appear to contain an empty element which moves into D. This might explain the obligatory absence of the definite article in structures containing a lexical subject. See also footnote 21.

[29] Thanks to Ian Roberts for helping me check these examples.

[30] Historically, though, there seem to have been some cases with NumP overt. In particular, some of the preverbal particles derive from pronouns, i.e. NumP in our account (*ml, be...*).

[31] The alternative, reducing all agreement to cases of incorporation, as proposed in Taraldsen (1992) does not seem exactly promising.

[32] I will not attempt to account for the agreement patterns under coordination as described in McCloskey and Hale (1984).

[33] I am aware of complications with the demonstrative pronoun *e seo* 'this', and not **seo teach* 'this house' but *an teach seo* 'the house this'. An analysis of this construction will take me too far afield.

[34] This is also true for Breton (cf. Stump (1989, p.438)).

[35] This generalization actually extends to complements: many languages cannot both license a possessor and a theme in their DPs, and can only license an internal argument if the genitive position is available.

[36] Here I assume that V movement somehow "activates" AgrO.

[37] Thanks to Arild Hestvik for the examples.

[38] Similar examples can be constructed for Dutch pronouns with PP modifiers. These pronouns have the distribution of lexical DPs; And, in particular, the pronoun cannot be a clitic form (*d'r* or *'m*):

(i) ik ken *d'r/ haar van boven niet goed.
I know her from upstais not well

This shows that when there is a clitic form, NumP to [Spec, DP] must take place, and when there is not, NumP to [Spec, DP] is optional. Note that this conclusion holds independently of the problem how these clitics should be analyzed: if they involve NumP movement, they must be preceded by NumP movement to [Spec, DP]; if they involve head movement, NumP movement to [Spec, DP] must be assumed as well.

[39] The star stands for the interpretation 'I don't like the person with long hair', not for a small clause construal 'I don't like him when he has long hair'.

[40] This account raises the question how object shift in Icelandic, where lexical DPs and pronominal DPs both shift, should be accounted for. I believe that this phenomena is much closer to scrambling in Dutch, which applies to both pronouns and specific DPs, whereas the pronoun/lexical DP asymmetry is closer to clitic placement, i.e. involves NumP movement to a pronominal head projection (Dutch actually has both processes). This implies that there is a distinction between AgrO (which could be accusative Case) and pronominal AgrO, which could be much closer to Sportiche's clitic projections (1995).

[41] The same is true for topicalization: (?)*all these people I really like*, **them all I really like*.

[42] Overt movement can be easily forced in Chomsky's (1992) theory where only morphological features can cause overt movement in the syntax (strong morphological features must be eliminated through checking in the overt syntax). English pronouns differ morphologically from English lexical Ns in carrying overt morphological Case distinctions. Pronominal movement could therefore be motivated as DP movement to the position where these features are checked. Asymmetries between pronominal DPs and lexical DPs would arise from the principle of Procrastinate (Chomsky (1992): lexical DPs don't move unless they are forced to). There are quite serious problems with Procrastinate, however. Consider for instance Chomsky's assumption that English accusative DPs do not (or do not optionally) move in the overt syntax. The distribution of DPs in the verb particle construction quite strongly argues in favour of optional DP movement of lexical DPs (Johnson (1991), Koopman (1991) among others). But if DPs can optionally move, the impossibility of structures like (i) cannot be explained:

(i) *with these people all

An account which forces movement of the entire pronominal DP thus runs into the problem that optional DP movement should be able to move the DP to the same positions as well.

[43] While pronouns move to [Spec, AgrS], [Spec, AgrO] and [Spec, AgrPP], [Spec, AgrN] is absent from this list, in contrast with Welsh. In fact, interesting problems arise with this position in DPs in English.

(i)
a. I spoke to the mother(SG) of all these children.
b. I spoke to the mothers of all these children.
c. ?*I spoke to all these children's mother.
d. (??) I spoke to all these children's mothers.
e. **I spoke to them/their all's mother(s).

While (ie) is crushingly bad, the source sentences (ic) and (id) are pretty unacceptable as well. It is unclear whether the interpretation in (ib) is available for (id): this sentence feels like a garden path, and judgments are accordingly. However, regardless of what explains (ic) and (id), if in addition, NumP must move to some higher Agr position, the ungrammaticality of (ie) is explained: the DP containing the pronoun is in [Spec, AgrN:] NumP cannot move to a higher Spec position, say [Spec, AgrO], or [Spec, AgrP] because it is not governed from the outside.

[44] It is often argued that English pronouns are clitics.

[45] D can be overt or covert in Welsh.

[46] Or maybe a silent Agr language and a silent NumP language, as Chinese/Japanese.

[47] If French finite verb do not occur in Agr, but lower, and the V in other Romance languages moves to Agr, it would follow French is not (subject) pro-drop.

[48] There are typically Agr asymmetries with overt object agreement. Overt object agreement is often restricted to specific DPs. This might follow if agreement is always and only triggered by NumP. If the [NumP, D] structure is a prerequisite for triggering of object agreement, you must belong to a particular class of DPs for NumP extraction to be possible. If the DP is lower than AgrO, there will be no agreement. If it is in a based generated A'-position, there will be agreement with a NumP in [Spec, Agr].

[49] Finally, the question arises what light this analysis sheds on clitic doubling constructions. In Welsh, pronouns are obligatorily doubled by Agr: this clitic doubling was described as NumP movement to [Spec, Agr] with D stranding This suggests a similar analysis for clitic doubling constructions quite generally. The problem in clitic doubling constructions is finding a source for the clitic. In clitic doubling constructions, the doubled DP must typically be precede by a dummy P (Spanish *a*, Hebrew *set*,...) Classical analyses of clitic doubling constructions (Jaeggli (1981), Borer (1984) for example) assume that the clitic absorbs Case, and that the function of the dummy P is to Case mark the DP. The analysis presented in this paper provides another way of looking at the appearance of P, however. In the absence of P, there is no source for NumP movement: the lexical NumP is within the DP. However, it could be proposed that the dummy P does exactly that: it creates an additional position from which NumP extraction is possible. The dummy P, creates an additional agreement projection, AgrP, in which an expletive NumP occurs, agreeing with NumP in the DP. It is the expletive NumP that moves to a higher [Spec,Agr], yielding clitic doubling. In other words, these dummy Ps are parasitic on P and project part of the PP structure; this allows the presence of an expletive NumP. This NumP can further move to [Spec, Agr], which is only possible if there is no C like type position dominating Agr, i.e. these dummy Ps project transparent PP small clauses (in that way they resemble Ds).

REFERENCES

Abney, S.: 1987, *The English Noun Phrase in its Sentential Aspect*, Doctoral Dissertation, MIT.

Anderson, S.: 1982, 'Where is Morphology', *Linguistic Inquiry* 12, 571 -612.

Awberry, G.: 1976, *The Syntax of Welsh: A Transformational View of the Passive*, Cambridge University Press, London.

Awberry, G.: 1990, 'Dialect Syntax: a neglected source for Welsh', in R. Hendrick ed.

Baker, M.. and K. Hale: 1990, 'Relativized Minimality and Pronoun Incorporation', *Linguistic Inquiry* 21, 289-297.

Borer, H.: 1984, *Parametric Syntax*, Foris Publications, Dordrecht.

Borsley R. and J. Stephens: 1989, 'Agreement and the position of subjects in Breton', *Natural Language and Linguistic Theory* 3, 407-427.

Cardinalleti, A.: 1993, 'On the Internal structure of pronominal DPs', ms. University of Venice.

Carstens, V.: 1991, *The Morphology and Syntax of Determiner Phrases in KiSwahili*. Doctoral Dissertation, UCLA.

Carstens, V.: 1993, *Deriving agreement*, ms, Cornell University.

Chomsky, N.: 1991, 'Some Notes on Economy of Derivation and Representation', in R. Freidin, ed., *Principles and Parameters in Comparative Grammar*, MIT Press, Cambridge.

Chomsky, N.: 1992, *Towards a Minialist Program for Linguistic Theory*, MIT Occasional Papers in Linguistics, 1.

Chung, S. and J. McCloskey: 1987, 'Government, barriers and small clauses in Modern Irish,' *Linguistic Inquiry* 18, 173-237.

Corver, N.: 1990, *The Syntax of Left Branch Extractions*, Doctoral Dissertation, Tilburg.

Crook, H.: 1990, 'The intemal structure of DPs in Bambara', MA thesis, UCLA

Grimshaw, J.: 1991, 'Extended Projections', ms. Brandeis University.

Hendrick, R.: 1989, *Anaphora in Celtic and Universal Grammar* Kluwer, Dordrecht.

Hendrick, R.: 1990, *The Syntax of Modern Celtic Languages*, San Diego: Academic Press, 27-79.

Hestvik, A.: 1990, *LF movement of Pronouns and the Computation of Binding Domains*, Doctoral Dissertation, Brandeis University.

Holmberg, A.: 1986, *Word Order and Syntactic Features in the Scandinavian Languages and English*, Doctoral Dissertation, Stockholm.

Jackendoff, R.: 1990, *Semantic Structures*, MIT Press. Cambridge, Mass.

Jaeggli, O.: 1981, *Topics in Romance Syntax*, Foris Publications.

Johnson, K.: 1991, 'Object Positions', *Natural Language and Linguistic Theory*, 9, 577-637.

Kayne, R.: 1985, 'Principles of Particle Constructions', In J. Gueron, et al. eds. *Grammatical Representation*, Foris Publications, Dordrecht.

Kayne, R.: 1994, *The Antisymmetry of Syntax*, Linguistic Inquiry Monograph Twenty-five, MIT Press, Cambridge.

Koopman, H. and D. Sportiche: 1989, 'Pronouns, Logical Variables and Logophoricity in Abe', *Linguistic Inquiry*, 20, 555- 589.

Koopman, H. and D. Sportiche: 1991, 'The Position of Subjects', Lingua 85.2/3.

Koopman, H.: 1984, *The Syntax of Verbs*, Foris Publications, Dordrecht.

Koopman, H.: 1988, 'The Structure of the VP in Dutch', paper presented in Geneva.

Koopman, H.: 1991, 'The Verb Particle Construction and the Syntax of PPs', ms. UCLA.

Koopman, H.: 1993, *The Structure of Dutch PPs*, ms. UCLA.

Koopman, H.: 1994, 'Licensing Heads', in N. Hornstein and D. Lightfoot, eds., *Verb Movement*, Cambridge University Press.

Koopman, H. and D. Sportiche: 1982, 'Variables and the Bijection Principle', *The Linguistic Review* 2,139-160.

McCloskey J. and K. Hale: 1984, 'On the syntax of person-number inflection in Modern Irish', *Natural Language and Linguistic Theory* 6, 143-189.

McCloskey, J.: 1992, 'On the Scope of Verb Movement in Irish', unpublished paper, UCSC.

Moritz, L.: 1993, 'DPs in Ncufie',In H. Koopman and M. Kural, eds, *The structure of Ncufie,* UCLA Working papers..

Nkemnji, M.: 1995, 'Heavy Pied-piping in Nweh,' Doctoral Dissertation, UCLA.

Postal, P.: 1969, 'On the so-called "pronouns"in English', In D. Reibel and S. Shane, eds. *Modern Studies in English*, Prentice-Hall, Englewood Cliffs NJ, 201-224.

Ritter, E.: 1991, 'Two Functional categories in Noun Phrases: Evidence from Modern Hebrew', ms. UQAM..

Rouveret, A.: 1990, 'X-bar Theory, Minimality and Barrierhood in Welsh', In R. Hendrick ed,

Rouveret, A.: 1991, 'Functional Categories and Agreement', In *The Linguistic Review*, 353-387

Sadler, L.: 1988, *Welsh Syntax. A Government-Binding Approach*, Croom H. Kent.

Sportiche, D.: 1995, 'Sketch of a Reductionist Approach to Syntactic Variations and Dependencies', In H. Campos and P. Kempchinsky, eds., *Evolution and Revolution in Romance Syntax*, Georgetown University Press.

Sportiche, D.: 1995a, 'Clitic Constructions', In L. Zaring and J. Rooryck, eds., *Phrase Structure and the Lexicon*, Kluwer Academic Publishers, Dordrecht.

Sportiche, D.: 1990, 'Movement, Agreement and Case', ms, UCLA.

Stowell, T.: 1991, 'Small Clause Restructuring', In R. Freidin, ed., *Principles and Parameters in Comparative Grammar*, MIT Press, Cambridge.

Stump, G.: 1984, 'Agreement versus Incorporation in Breton', *Natural Language and Linguistic Theory* 2, 289-348.

Stump, G.: 1989, 'Further Remarks on Breton Agreement', *Natural Language and Linguistic Theory* 3, 429-472.

Szabolci, A.: 1987, 'Functional Categories in the Noun Phrase', In I. Kenesei, ed., *Hungarian* vol. 2. Budapest: Jate Szeged.

Taraldsen, T.: 1992, 'Pronouns and agreement', GLOW newsletter.

Valois, D.: 1991, *The Internal Syntax of DP*, Doctoral Dissertation, UCLA.

Watkins, T. A.: 1972, 'The Welsh personal pronoun', *Word* 28, 146-165.

Williams, S. J.: 1980, *A Welsh Grammar*, University of Wales Press, Cardiff.

SPANISH PASSIVIZED DATIVES: THE RELEVANCE OF MISANALYSIS

Mario Montalbetti
University of Arizona, Tucson

According to XVIIth century puritan thought (cf. Taylor 1989), God doesn't care much for Direct Objects, meaning: *what* you do is not as important as *how* you do it. Thus, Joseph Hall's famous line: "God loveth adverbs". Humoring the puritan's grammatical metaphor we can speculate that God's second love falls on Indirect Objects, the dative beneficiaries of one's actions. And it is of God's second love that I'll write here; in particular about certain seemingly ill-behaved Spanish Indirect Objects that undergo Passivization. Towards the end, I will be recommending a treatment that has the unfortunate consequence of removing them altogether from God's affection.

It seems clear from the outset that there are theoretical tensions between the notions of "dative" (and in some cases "intransitivity") and "passivization"; and although the richness of the latter notion is a tribute to the ingenious articulation of such tensions, languages like Spanish are still recalcitrant guardians of the tradition that relates transitivity, Direct Objects and Passivization in exclusive terms. Put in other words, Spanish Direct Objects may passivize (1b), not so Indirect Objects (1c):

(1) a. Juan le vendió una casa a María
John cl:dat sold a house to Mary
'John sold a house to Mary'

b. Una casa le fue vendida a María
A house cl:dat was sold to Mary

c. *María fue vendida (P) una case

where P stands for any preposition you like, assuming that Spanish (argument) clauses may need Case, a point to which I will return in a moment.

And yet, under certain circumstances which I will detail in a moment, Indirect Objects can undergo Passivization (in several dialects of Spanish; I'll present evidence from my own, Peruvian Spanish). A paradigmatic case is the following:

(5) a. Juan le prohibió leer el libro a María

K. Johnson and I. Roberts (eds.), Beyond Principles and Parameters, 133-144.

Juan cl:dat prohibited to-read the book to María
'John prohibited Mary to read the book'

b. Leer el libro le fue prohibido a María
to-read the book cl:dat was prohibited to Mary

c. María fue prohibida de leer el libro
Mary was prohibited of to-read the book

A note on word order: it has been noted (cf. e.g. Plann (1986)) that a sentential Direct Object usually follows the Indirect Object, such that *Juan le prohibitó a María leer el libro* would be a more "neutral" S-structure order. As we'll see later this may be of some relevance. The order in (5a) is the logico-grammatical order of citing the Direct Object before the Indirect one. Both orderings are grammatical.

Now (5a) is an active sentence containing both a Direct Object *leer el libro* and an Indirect Object *a María*. Note that the Direct Object is an infinitival clause. The clitic *le* is doubling the Indirect Object *a María*.

Let me get out of the way the reasonable suspicion that the Peruvian dialect is a 'leista' dialect, a circumstance that would vitiate everything I've got to say; it is not. It never uses the dative clitic with Direct Objects as, for example, in some Peninsular dialects where *Juan le mató* is accepted.

(5b) is the standard passive version of (5a), where the sentential Direct Object has raised to Subject position (i.e. Spec of IP). The *by*-phrase is optional.

(5c) is the exceptional specimen. In it, the Indirect Object *a María* has raised to Subject (dropping the preposition *a*) and the Direct Object *leer el libro* appears now as the oblique Object of the preposition *de*. From a strictly impressionistic point of view it is rather clear what has happened. The raising of the Indirect Object seems to have been triggered by Case theoretic reasons and it has generated Case theoretic consequences; in particular, it has left the Direct Object Caseless. This is of some relevance. It has been noted by Susan Plann (1986) that Spanish Clauses may be Case marked. Although her conclusion was framed within a discussion of Stowell's Case Resistance Principle, we can adopt a stronger version of Plann's conclusion and suggest that Spanish clauses must be Case marked (whenever they are arguments; hence the difference between a Relative Clause and a Complex NP). If so, then in (5c) the preposition *de* has been inserted as a dummy Case-marker to save the construction. Of course, (5c) is perfectly grammatical in the dialect of Spanish under discussion.

As I've said, this curious structure occurs only under four (rather strict) conditions which I now turn to.

A) It only occurs with a certain set of verbs like

(6) {prohibir, permitir, impedir, ordenar,...}

all of which can be loosely characterized as "performative". Following Searle and others, let's call a verb V "performative" if the utterance of V itself brings into existence the very nonlinguistic action denoted by V. This means that in the case of a verb like *impedir* ('to impede') the passive version found in (7c) must be interpreted as a legal impediment, not a physical one. This indeed is the case. (Note for example, **María fue impedida de leer el libro por la falta de luz* where the reason of the impediment is the absence of light). The relevant point here is that the set of verbs that allow for this construction seem to form some natural class - even if the cover term "perfomative" is not the appropriate one.

B) The Direct Object must be a sentential and it must be untensed (i.e. it must be an infinitival clause).

(7) a. Juan le impidió [ir al cine] a María
Juan cl:dat impeded [to-go to-the movies] to María

b. [Ir al cine] le fue impedido a María

c. María fue impedida de [ir al cine]

(8) a. Juan le impidió [que vaya al cine] a María
Juan cl:dat impeded [that goes+3S to-the movies] to María

b. [Que vaya al cine] le fue impedido a María

c. *María fue impedida de [que vaya al cine]

(9) a. Juan le prohibió ese libro a María
John cl:dat prohibited that book to Mary

b. Ese libro le fue prohibido a María

c. *María fue prohibida de ese libro

In (8c) the Direct Object is tensed and in (9c) it is non-sentential, hence the ungrammatical results.

C) The PRO Subject of the Sentential Direct Object must be controlled by the Indirect Object.

(10) a. Juan le permitió [PRO$_1$ comer un chocolate] a María$_1$
Juan cl:dat allowed [PRO to-eat a chocolate] to Mary
'John allowed Mary to eat a chocolate'

b. [PRO$_1$ comer un chocolate] le fue permitido a María$_1$

c. Marí1 fue permitida de [PRO$_1$ comer un chocolate]

(11) a. Juan$_1$ le prometió [PRO$_1$ comer un chocolate] a María
Juan cl:dat promised [PRO to-eat a chocolate] to Mary
'John promised Mary to eat a chocolate'

b. ?[PRO$_1$ comer un chocolate] le fue prometido a María$_2$

b'. [PRO$_1$ darle un chocolate] (nunca) le fue prometido a María$_2$

c. *María$_1$ fue prometida de [PRO$_2$ comer un chocolate]

In (10c) Passivization of the Indirect Object is successful given that *permitir* ('to-allow') is not a Subject Control verb. (11c) on the other hand contains *prometer* ('to promise') which is a Subject Control verb and Passivization fails. Notice that with *prometer* even the Passivization of the Direct Object is dubious, but it improves within certain contexts as Bresnan (1982) has shown, and depending on whether the Control is obligatory or optional. In any case (11c) is completely out.

Other examples show the same pattern:

(12) a. Juan$_1$ le$_2$ confesó [PRO$_1$ serle$_2$ infiel] a María$_2$
Juan cl:dat confessed [PRO to-be-cl: dat unfaithful] to María

b. [PRO$_1$ serle infiel] le fue confesado a María$_2$

c. *María$_2$ fue confesada de [PRO$_1$ serle infiel]

A final property of this construction is

D) The Passivized dative must move (to Spec of IP).

(13) a. María fue prohibida de [PRO leer el libro]
b. *fue prohibida María de [PRO leer el libro]

a surprising finding given that in normal Passivization of the Direct Object it can stay in place:

(14) le fue prohibido [PRO leer el libro] a María

[The version *fue prohibida de leer el libro María* is OK only with a pause before *María*].

A rather clear picture thus emerges, at least if we consider the first three properties: Passivization of the Indirect Object is possible in this dialect (only) when the Direct Object is an Infinitival Clause whose PRO-Subject is controlled by the Indirect Object. Note again that the expected Case consequences follow straightforwardly.

Now of course, several questions arise; among others, what is the relative structural positions of both Direct and Indirect Objects, beyond, that is, the purely linear representation offered in (15).

(15) [NP V [PRO_1 ...] [PP ...NP_1]]
D. Object I.Object

It seems that, at least a priori, any representation that takes both Direct and Indirect Objects as separate entities will run into problems to explain the Passivization patterns just reported, regardless of whether we adopt Extended Theory structures or any of the Larsonian variations available, including Bowers' predication constructions. Again, whatever the structure, it will only be the Direct Object that will passivize. Yet another possibility is to assume the interchangeability of Direct and Indirect Objects. That would mean that the source of (16a) would be (16b)

(16) a. María fue permitida de ir al cine
Mary was allowed of to-go to-the movies

b. Juan permitió a María a ir al cine
John allowed Mary to to-go to-the movies

where *María* somehow appears as the Direct Object and the infinitival clause as an Indirect Object. But then why doesn't the dummy *de* appear in the passive version (16b) as in the previous cases? Still, this may be a feasible alternative.

The approach which I would like to try here takes a different route: it starts by looking at certain structures in which Case marking and Grammatical function do not follow the expected pattern of Subject/Nominative, Direct-Object/Accusative, Indirect-Object/Dative. In many cases what these structures have in common is a representation in which a sentential construction is the unitary Object of a Verb. (Consider, for example, ECM structures). Hence, what I'd like to suggest is that the apparent Indirect Object of the Spanish sentences under study is a Dative yes (in terms of Case marking) but it should be interpreted as the Subject of the Infinitival Direct Object clause. Thus, the representation I'd like to work with will look something like the following:

(17) Juan le prohibió [a-María leer el libro]
Juan cl:dat prohibited [to-Mary to-read the book]

If so, then the basic properties just outlined would follow more or less straightforwardly and it would be no coincidence now that this special type of Passivization occurs only when the Indirect Object Controls the Subject position of the Infinitival Clause. In fact, under this proposal, it doesn't "control" it but instead should be regarded as the very Subject of the said Clause. Furthermore, under this view it is not the case that Indirect Objects Passivize but that dative-marked NPs do. Of course, a dative-marked NP need not occur in an Indirect Object position.

Motivation for this move can be drawn from the analysis of a similar construction, the so-called perception verb structure. Consider a typical case like the following:

(18) Juan vio a María correr
Juan saw María to-run
'John saw Mary run'

A quick comparative analysis between English and Spanish yields two important differences, perhaps related: (a) as can be seen in (18) the Spanish version contains a full-blown infinitival complement while the English gloss contains what Jim Higginbotham (1983) originally called an "Unsupported Clause" (i.e. one containing a "naked infinitive"); (b) the second difference is that while *María* can be Passivized in the Spanish version, *Mary* can't in the equivalent English construction:

(19) a. María fue vista correr
b *Mary was seen run

Regardless of these differences I believe it is still possible to assign the same structure to both constructions, one in which the perception verb takes a clausal complement (whose Subject is *María/Mary*) as in (20a,b).

(20) a. Juan vio [a-María corder]
b. John saw [Mary run]

Different from the Passivized dative construction the Case marking on this Subject is accusative, as (21) shows:

(21) a. Juan la vio correr
John cl:acc saw to-run

b. John saw her run

Notice that a-priori there is another structure readily available, namely:

(22) a. Juan vio a María$_1$ [PRO$_1$ correr]
b. John saw Mary$_1$ [PRO$_1$ run]

Recall that this very dual complement alternative was the form we adopted to introduce our Passivized datives.

However, there are various arguments in favor of the clausal complement hypothesis for perception verbs. An immediate syntactic proof is readily available for English through well known examples like *I saw it rain*, where the pleonastic *it* can hardly be regarded as a Direct Object. Additionally some semantical generalizations noted by Barwise (1981) and discussed in Higginbotham (1983) tend to support this analysis (i.e. the representations in (20)). They are:

(23) a. Veridicality: the Truth of the following conditional: If John sees Mary run → Mary runs. (Si Juan ve a María correr → María corre)

b. Substitutivity: the unsupported complement of a perception verb is a transparent domain open to substitution.

c. Exportability: existential quantifiers taking widest possible scope within the unsupported complement of a perception verb are exportable. Thus, conditionals like the following are true:

If John sees somebody leave → there is somebody whom John sees leave.
(Si Juan ve a alquien salir → hay alquien a quien Juan ve salir)

Two out of these three generalizations also apply to our Indirect Object constructions. The first one, veridicality, clearly doesn't probably because perception verbs are epistemic verbs which the "performative"-verbs used in our constructions aren't. However, substitutivity and exportability are properties also found in our constructions. For example:

(24) Si Juan le prohibió [a-María leer un libro] → Juan le prohibió [a-la hija de Susana leer un libro], given that María = la hija de Susana.

(25) Si Juan le prohibió [a-alquien leer un libro] → hay alguien a quien Juan le prohibió leer un libro.

But perhaps more importantly some nominalization facts seem to support the view that in both cases, in perception verbs and in the Passivized dative structures, we have a unitary complement of the verb. Consider specifically, the contrast between the following sentences:

(26) a. Juan le prohibió [a-María leer el libro]
Juan cl:dat prohibited [to-Mary to-read the book]

b. Juan ordenó la prohibición
John ordered the prohibition

c. La prohibición = que María lea el libro
The prohibition = that Mary read the book

d. Juan prohibió [que María lea el libro]
John prohibited [that Mary read the book]

e. Juan prohibió [que María lea el libro]
John prohibited [that Mary read the book]

(27) a. Juan le entregó el libro a María
John cl:dat delivered the book

b. Juan ordenó la entrega
John ordered the delivery

c. La entrega = el libro
The delivery = the book

The nominalization expressed in (26b) covers the contents of the whole unitary complement clause. Not so with (27) where the nominalization restricts itself to the Direct Object NP leaving out the Indirect Object. Furthermore (26a) could be paraphrased quite closely by (26d,e).

Perhaps these contrasts can be derived from the fact that (28a) doesn't entail (28b) but (29a) entails (29b). This contrast in turn may be derived from a logical form for (28a) like the one that appears in (28c).

(28) a. Juan le prohibió a María leer el libro
b. *→Juan prohibit leer el libro
c. ∃e[prohibición (e) & Subject (e, Juan) & ∃e' [leer(e') & Object(e, e') & Subject(e', María) & Object (e', el libro)]]

(29) a. Juan le entregó el libro a María
b. → Juan entregó el libro
c. ∃e[entrega(e) & Subject(e, Juan) & Object(e, el libro) & Indirect Object(e, María)]

But perhaps the most striking parallel between perception verbs and Passivized dative constructions is that they take the same type of (infinitival) complement; indeed compare:

(30) a. Juan la vio a María correr
Juan cl:acc saw María to-run

b. Juan le prohibió a María correr
Juan cl:dat prohibited to Mary to-run

I have doubled the Direct Object in (27b) to stress the parallel. [I haven't yet checked the obvious question whether the Passivization of Indirect Objects is a property of Clitic Doubling dialects. The Peruvian Dialect is a doubling dialect]. Note that the only difference is that in one instance *María* is marked Accusative and in the other Dative.

The parallel is all the more striking when we observe that both constructions observe the fourth property (D) mentioned above in the sense that under Passivization movement is required:

(31) a. María fue vista correr
Mary was seen to-run

b. *Fue vista María correr

(32) a. María fue prohibida de correr
Mary was prohibited of to-run

b. *Fue prohibida María de correr

Before going back to our original examples, let me solve a possible problem for the structural parallel we are drawing between perception verb constructions and Passivized dative structures. We have said that one of the properties of perception verbs in Spanish is that the accusative Subject of their complement clause can Passivize (cf.(18)). Why isn't it the case then that after Passivization a dummy Case-marker is inserted to assign Case to the infinitival clause, resulting in something like (33a)

(33) a *María fue vista de correr
b. María fue prohibida de correr

which contrasts with (33b). However, notice that even before Passivization both structures differ. The source for (18) is (34a) not (34b):

(34) a. Juan vio a María correr
b. *Juan vio de a María correr

This might follow from the possibility that even though the complement is taken to be unitary, finer distinctions still obtain. For example, at some level, the infinitival complement of perception verbs may be taken to be an adjunct (and hence needs not Case; if only argument clauses require Case in Spanish). In the Passivized dative construction the infinitival clause is taken to be an argument and hence requires to be

Case marked. Alternatively, maybe the infinitival verb in perception verb complements is truly naked, even in Spanish, and hence could avoid Case requirements.

Let us regroup now. We have seen that certain verbs that take an infinitival clause as a Direct Object allow their Indirect Objects to Passivize provided that the Indirect Object controls the PRO-Subject of the infinitival clause. We have suggested a different analysis from the purely impressionistic one by assuming that the Indirect Object is in fact the dative Subject of the infinitival clause. We have drawn support from the better understood analysis of perception verb structures, from the striking similarity of complements that both perception verbs and our "performative" verbs can take (as seen in (30a,b) and also from their behavior under Passivization). We are thus left with a representation such as the following:

(35) NP V [NP:dat V:inf]

Noting the similarity with the perception verb structure:

(36) NP V [NP:acc V:inf]

There is one more fact worth noting. Indeed, we have been dealing with structures in which the Direct Object is Sentential and the Indirect Object is not. What about the reverse case, when the Direct Object is an NP and the Indirect Object is sentential (assuming the same Control facts). Consider the following examples:

(37) a. Juan la forzó a María a [leer el libro]
John cl:acc forced María to [to-read the book]

b. María fue forzada a [leer el libro]

c. [leer el libro] le fue forzado a María

Notice (37c) which is indeed the Passivization of an Indirect Object again, only that this time it is sentential. However, here we have a completely different structure. Notice that in (37a) the Indirect Object is preceded by a Preposition (*a*) hence making it improbable to assume that the verb is taking a unitary complement like in (38):

(38) Juan la forzó [a María a leer el libro]

Furthermore notice that the same conditions and properties that held with respect to the original cases of dative Passivization do not hold here. For example: the Passivized sentential dative in (35c) could have stayed in place:

(39) Le fue forzado [leer el libro] a María
(or: le fue forzado a María [leer el libro])

Moreover, there is no straightforward paraphrasis with a unitary tensed complement as was possible in (26e). Compare:

(40) *Juan forzó (a) [que María lea el libro]

If (40) means anything it certainly doesn't mean that John forced Mary to read the book.

So in this instance we must assume a dual complement structure and despite some superficial similarities suggest that different processes are taking place here. Indeed, a suggestion which we regarded as infeasible for the original Passivized dative cases could well be rescued to deal with this second type of phenomenon. In other words, what examples like (37) may illustrate is some degree of interchangeability of the Direct and Indirect Objects, but I'm not prepared to say much more at the moment.

It would be nice to end with some sort of conclusion especially because conclusions give the feeling that one is in control of both theory and data. But in this case I'm not. And I'm sure that I've aggravated things by talking about a rather small dialect and by employing the mathematician's strategy of assimilating one problem to another whose solution is better understood. That I've done by suggesting that in essence, perception verb structures and the Passivized dative structures I've been dealing with have the same basic properties. What remains, of course, is to answer why only a small dialect of Spanish has exercised this option. So let me venture a final hypothesis. I submit that the Passivized dative construction of my dialect is a case of the speaker's honest analytical mistake. Indeed, we have come to analyze the Passivized dative construction as a perception verb structure. Again the question is why. There are several contributing factors, but there are two key ones: first, notice that both the "dummy" preposition *a* that introduces certain animate, definite Direct Objects and the real preposition *a* that introduces the Indirect Object have one and the same phonological form. This coincidence is compounded in the case of our construction because the dative Object can be Passivized only when the main verb selects complement clauses whose Subjects are the dative Objects in question. Such set of verbs will uniformly take animate Indirect Objects.

The second key factor is that even though our dialect is not a "leista" dialect and we clearly use *le* for Indirect Objects and *lo/la* for Direct Objects, the agreement relation between the dative clitic and the Indirect Object is extremely weak in terms of agreement (as I have showed several years ago (Montalbetti (1981))). In fact, not only no gender agreement obtains between the dative clitic and a doubled Indirect Object but number agreement is many times lost, as the grammaticality of examples like (41) shows:

(41)	Juan le	dio	un caramelo	a los niños
	Juan cl:dat:Sing	gave	a candy	to the kids

Combining these two factors this is what the structure of perception verb structures and passivized dative constructions will look like:

(42) [NP cl V [a-NP V:inf...]

In the case of the perception verb structure the cl=acc and the *a* is a dummy preposition; in the case of the Passivized dative constructions the cl=dat and the *a* is a real preposition (introducing an Indirect Object).

Although I have called this (mis)analysis a case of an honest mistake, the resulting Passivized dative construction reveals a certain feeling of guilt because it recognizes the need to insert a Case-marking preposition (*de*) when the dative Object has moved to Spec of IP, hence admitting after all a small difference with the perception verb structure. If linguistic fiction is still a permissible activity I would believe that the next step in this peculiar construction will be the dropping of this preposition. But, of course, we'll have to wait and see.

REFERENCES

Barwise, J.: 1981, 'Scenes and Other Situations', *The Journal of Philosophy*, LXXVIII.

Bresnan, J.: 1982, 'Control and complementation', *Linguistic Inquiry*, 13.3.

Higginbotham, J.: 1983, 'The Logic of Perceptual Reports: An Extensional Alternative to Situation Semantics', *The Journal of Philosophy*, LXXX.

Plann, S.: 1986, 'Case-Marking Clauses in Spanish: Evidence against the Case Resistance Principle', *Linguistic Inquiry*, 17, 336-346.

Taylor, C.: 1989, *Sources of the Self*, Harvard University Press, Cambridge.

THE SELECTION PROPERTIES OF ASPECTUAL VERBS*

Anne Rochette
Université du Québec à Montréal

In the generative literature, it has generally been assumed that aspectual verbs like *begin* show mixed behaviour with respect to the distinction established between raising and control predicates. Although these verbs clearly behave as raising predicates, as shown by the paradigm sentences in (1) first pointed out by Perlmutter (1970), the fact that they can also occur in simple transitive structures, as shown in (2), has led many syntacticians to posit the existence of two subcategorization frames for these verbs.[1]

(1) a. There began to be a commotion.
b. It began to rain.
c. Headway began to be made toward a solution.
d. The noise began to annoy Joe.

(2) John began the letter.

The thesis that will be defended in this paper is that all occurrences of aspectual verbs should be analyzed as raising predicates, including the occurrences in simple transitive structures such as (2). It will be shown that the different structures in which aspectual verbs can appear all exhibit selectional restrictions with respect to the complement introduced by these verbs, whether this complement is a non-finite complement or a nominal complement. More precisely, it will be argued that aspectual verbs select the semantic category *process* and that this semantic category can be realized either as a verbal projection or as a nominal projection. It will also be argued that the surface subject of aspectual verbs is always selected by the embedded predicate even when this predicate is not lexically realized as in the case of sentence (2).

This paper will be organized as follows. In the first section, I will outline the general framework of sentential complementation that underlies the present study. The second section will focus on the syntactic and semantic properties of the constructions in which aspectual verbs introduce non-finite complements. I will first briefly review the arguments found in the literature in favour of treating aspectual verbs as either raising predicates or control predicates. It will be shown that the facts that militate in favour of a control analysis of these verbs can be independently explained and therefore do not undermine the raising analysis. I will conclude this section by presenting other properties of the non-finite constructions that indicate that aspectual verbs show a behaviour close to that of auxiliary verbs. The third section will be devoted to the study

K. Johnson and I. Roberts (eds.), Beyond Principles and Parameters, 145-165.

of the selectional restrictions imposed by aspectual verbs with respect to the event-type of their non-finite complements. It will be shown that aspectual verbs can introduce activities or accomplishments but that they do not generally admit states or achievements. It will be proposed that activities and accomplishments form a natural aspectual class which can be characterized in terms of the semantic category *process*. Finally, in the fourth section, it will be shown that selectional restrictions are also operative in the case of the transitive structures in which aspectual verbs appear to select a nominal direct object and it will be argued that these constructions are also instances of raising structures.

1. SYNTACTIC & SEMANTIC SELECTION OF SENTENTIAL COMPLEMENTS

Within the framework of generative grammar, selection has traditionally been handled in terms of subcategorization frames which encode the syntactic categorial type of the (internal) arguments selected by a given predicate (Chomsky 1965). Grimshaw (1979) has however shown that it is also necessary to postulate the existence of a mechanism of semantic selection to account for the interpretation of certain structures which receive different interpretations according to the matrix verb that introduce them.[2]

In Rochette (1988), I developed a general approach to sentential complementation articulated in terms of the mechanism of semantic selection.[3] I proposed to distinguish three major semantic categories of complements: *action*, *event* and *proposition*. These semantic categories are respectively selected by predicates belonging to three distinct semantic classes of predicates: the effective, emotive and propositional predicates.
This classification of main predicates, which is illustrated in (3), is borrowed from the study of Long (1974). Long establishes a first distinction between effective and reflective predicates. Effective predicates describe a subject's relationship - whether causal, potential or other - to the performance of an action whereas reflective predicates express a (human) subject's judgment concerning a proposition or an event. Long also establishes a further distinction between two types of reflective predicates: predicates of the emotive type which express judgments of "personal relevance" and predicates of the propositional type which express judgments of "truth value".

(3)

effective		reflective	
begin	emotive		propositional
	wish		*believe*

Interestingly, this trichotomy in the semantic classes of predicates and in the semantic categories that they select also generally reflects a trichotomy in the types of sentential complements selected by these predicates. Predicates of the effective class can only appear with infinitive or gerundive complements whereas predicates of the reflective type may appear with infinitive as well as with tensed complements. Furthermore, in the Romance languages, predicates of the emotive type select subjunctive complements while predicates of the propositional type select indicative complements.[4] The following French examples illustrate these selectional restrictions.

(4) a. Jean commence à écrire son livre.
'John begins to write his book.'

b. *Jean commence qu'il écrit/écrive son livre.
'*John begins that he writes/write his book.'

(5) a. Jean souhaite partir.
'John wishes to leave.'

b. Jean souhaite que Marie parte demain.
'John wishes that Mary leave tomorrow.'

(6) a. Jean croit avoir compris l'explication.
'John believes that he has understood the explanation.'

b. Jean croit que Marie a compris l'explication.
'John believes that Mary has understood the explanation.'

In order to account for the selection of these different types of sentential complements as well as for their varying syntactic properties, I argued that the three semantic classes of main predicates select three distinct semantic categories which are structurally realized as projections of distinct syntactic categories as well. The effective predicates select the semantic category *action* which is structurally realized as a projection of the category V, thereby accounting for the lack of tensed complements with these predicates. The emotive predicates select the semantic category *event* which is structurally realized as a projection of the category Infl; the choice of the subjunctive mood follows from the selection of Infl as does also the absence of a distinct Tense operator for the embedded clause. The propositional predicates select the semantic category *proposition* which is structurally realized as a projection of the category Comp. According to the framework just outlined, aspectual verbs would belong to the class of effective predicates since they do not express any judgment on the part of the subject. It is not clear however that they truly describe a subject's relationship to the performance of an action in the same way than a verb like *try* does. They rather function as aspectual modifiers with respect to the action described by the embedded predicate and its arguments, including the external argument which surfaces as the grammatical subject of the aspectual verb. I will however continue to treat them as belonging to the effective class.[5]

Aspectual verbs would thus semantically select the category *action*. Originally, the choice of the term *action* was meant to capture the fact that, in general, effective predicates do not easily allow stative verbs as their complements. The semantic category *action* would therefore contrast with that of *state* and would presumably comprise three of the four aspectual classes distinguished by Vendler (1967), namely the accomplishments, the activities and the achievements. In section 3, however, we will see that this opposition between the categories *action* and *state* does not account for the whole range of selectional restrictions found with aspectual verbs and that it is necessary to postulate the existence of a semantic category that distinguishes accomplishments and activities from achievements and states. But before turning to this issue, I would first like to examine some of the syntactic and semantic properties of the constructions in which aspectual verbs introduce non-finite complements.

2. THE NON-FINITE COMPLEMENTS OF ASPECTUAL VERBS

The constructions in which aspectual verbs introduce non-finite complements present a number of syntactic and semantic properties that indicate that the embedded predicate is the main predicate of the construction with respect to the selection of the arguments and with respect to the interpretation of the sentences. In this section, I will first examine the status of these verbs with regard to the distinction made between raising and control predicates. I will then review some other semantic and syntactic properties associated with these non-finite complements that all point to a monoclausal analysis of these structures.

2.1. Raising vs. control

It is a well-known fact that aspectual verbs show properties characteristic of raising verbs. I repeat Perlmutter's examples given in (1) above, which all illustrate properties associated with raising structures.

(7) a. There began to be a commotion.
b. It began to rain.
c. Headway began to be made toward a solution.
d. The noise began to annoy Joe.

The four examples given in (7) show that aspectual verbs do not impose selectional restrictions on their surface subject and that the subject is rather selected by the embedded predicate.

Aspectual verbs however appear to constitute a hybrid class with respect to the distinction made between raising and control predicates in the sense that they show properties that are diagnostic of both classes. On the basis of examples such as the following, Perlmutter (1970) argues that aspectual verbs also enter into control structures since, in these examples, they appear to select an external argument.

(8) a. I tried to begin to work.
b. I forced Tom to begin to work.
c. Begin to work.
d. Sam began the job.
e. Mark began enthusiastically, but he got tired by noon.

The first two examples in (8) involve an embedding under typical control verbs such as *try* and *force*. This type of embedding was originally taken as a test for the distinction between raising and control, the argument being that control verbs require that the subject of their embedded infinitivals be a deep-structure-like subject (Perlmutter 1970:111-112). This argument however does not easily carry over under an analysis in terms of PRO subjects. Without any specific constraint to the contrary, it is perfectly possible to embed a raising predicate under a control one; the PRO surface subject of the raising predicate being simply derived from a deeper embedded position, as shown in (9).

(9) a. I_1 tried [PRO_1 to begin [t_1 to work]]
    b. I forced Tom_1 [PRO_1 to begin [t_1 to work]]

The example in (8c) is also questionable as an argument showing the control status of a predicate. This argument is based on the fact that a prototypical raising predicate such as *seem* appears to disallow the imperative, a fact which was taken to follow from the raising nature of *seem*. It is however well-known that the formation of the imperative is subject to an aspectual constraint which affects stative predicates. It seems to me that the oddness of the imperative with *seem* could well be attributed to that aspectual constraint rather than to the raising nature of the predicate. The fact that *begin* can be used in the imperative would therefore not constitute an argument in favour of a control status of the verb.

The examples in (8d-e) appear to indicate that *begin* must be able to select its surface subject since there is no embedded predicate here that could be held responsible for the selection of the subject. In section 4, however, it will be argued that the surface subject in these sentences is selected by the understood *process* predicate selected by the aspectual verb and that these two examples therefore also constitute instances of a raising construction.

Lamiroy (1987) also notices that the ungrammaticality of examples such as the following is unexpected under a raising analysis of aspectual verbs since these verbs should then freely allow raising of the expletive subject from the embedded complement.

(10) a. *Il continue à falloir partir.
    'It continues to be necessary to leave.'

    b. *Il commence à s'agir de travailler.
    'It begins to be necessary to work.'

These examples would indeed appear to indicate that the aspectual verb establishes a form of selection on its grammatical subject since the presence of the expletive *il* raised from the embedded infinitive gives results contrary to those obtained with raising verbs like *seem*.

It appears, however, that the ungrammaticality of these examples can probably be attributed to other factors that do not necessarily imply that we must analyze these occurrences of aspectual verbs as control verbs. In the next section, it will be shown that aspectual verbs do not generally introduce stative predicates or achievements. It seems to me that the existence of such selectional restrictions can be held responsible for the nonoccurrence of predicates like *falloir* and *s'agir* 'be necessary', which clearly show a stative character. Also, these two predicates belong to the class of propositional predicates and therefore introduce a proposition which appears to give to the whole complement the status of a proposition. The propositional status of the complement is obviously incompatible with the selection requirements of aspectual verbs.[6]

In a language like French, a further argument in favour of the raising status of aspectual verbs is provided by Ruwet (1972) who shows that the possibility for the adnominal pronoun *en*, which corresponds to a complement of the surface subject, to appear cliticized onto the embedded verb follows from the fact that the matrix subject originates in the embedded complement.

(11) a. La solution DU PROBLÈME commence à être connue.
'The solution of the problem begins to be known.'

b. La solution commence à EN être connue.

Ruwet (1983) however points out that the possibility for the adnominal pronoun *en* to appear attached to the embedded verb is not as free as it first seemed to be since example (12b) is less acceptable than (11b).

(12) a. L'auteur DU PROJET commence à être connu.
'The author of the project begins to be well-known.'

b. ?L'auteur commence à EN être connu.

Ruwet shows that the animate or inanimate character of the subject plays an important role in the contrast illustrated by these examples. Lamiroy (1987) also gives the following examples to illustrate this constraint.

(13) a. Le chef DU GOUVERNEMENT continue à être répressif.
'The head of the government continues to be repressive.'

b. ?*Le chef continue à EN être répressif.

In this case, however, the presence of the adnominal pronoun *en* is also not very acceptable in the following simple sentence involving the same elements.

(14) ?*Le chef en est répressif.

Ruwet and Lamiroy both suggest that the presence of an animate subject in these examples must be correlated with the notion of control verb and that of an inanimate subject with the notion of raising verb. It seems to me, however, that the fact that the placement of *en* obeys some kind of inanimacy constraint must be dealt with independently in the grammar, as shown by the marginal status of (14), and therefore does not undermine the raising analysis of these verbs.

To summarize, with respect to the non-finite constructions introduced by aspectual verbs, the evidence provided thus far clearly indicates that aspectual verbs behave like raising predicates. The raising status of aspectual verbs does however raise the question of the precise syntactic structure that should be associated with these non-finite complements. It is generally assumed that raising verbs such as *seem* introduce IP complements and that the absence of a CP structure with these verbs accounts both for the absence of control and for the fact that raising of the embedded subject does not violate any grammatical principles. Auxiliary verbs are also generally analyzed as raising elements although their "complements" are not analyzed as IP complements but rather as VP complements. Under the current approach whereby subjects are generated within the verbal projection, a VP analysis of the non-finite complements of aspectual verbs can be envisioned as an alternative to an IP analysis. But before addressing this issue any further, I would first like to review some further properties associated with these structures that may shed some light on the status of these non-finite complements.

2.2. Semantic and syntactic properties of the non-finite constructions

The interpretation of the constructions in which aspectual verbs introduce non-finite complements shows these verbs do not constitute autonomous events. In fact, semantically, they simply function as aspectual modifiers with respect to the event described by the non-finite complement. For instance, the interpretation of a sentence like (15a) does not involve two distinct events constituted of *begin* and *read this book* but rather denotes a single event of *begin to read.* In contrast, the interpretation of a sentence like (15b) involves two events: a first event constituted of the matrix predicate *wish* and its arguments and a second "unrealized" event constituted of the complement *read this book.*

(15) a. John begins to read this book.
b. John wishes to read this book.

The fact that it is impossible to use distinct time reference adverbials in constructions with aspectual verbs also indicates that these structures involve a single event. Again, the contrast between verbs like *begin* and *wish* is revealing.[7]

(16) a. *Yesterday, John began to read this book today.
b. Yesterday, John wished to read this book today.

Thus, semantically, aspectual verbs essentially appear to play a role of modifier with respect to the event denoted by the embedded predicate and its arguments. I therefore conclude that these structures involve single propositions at the level of semantic representation.

From a syntactic perspective, the constructions in which aspectual verbs occur also present a number of properties that are characteristic of monoclausal structures. For instance, the non-finite complements of aspectual verbs do not behave as objects of these verbs in that they cannot be directly questioned. To the question given in (17a), where *what* refers to the object of the verb *begin*, only the answer given in (17b), which involves a direct object noun phrase, is acceptable; the answer in (17c), which involves an infinitival complement, is not truly acceptable. In fact, it is necessary to have recourse to a "resumptive verb" like *do* to question the infinitival complement of verbs like *begin*, as shown by the question given in (17d) to which can correspond the answer (17c).

(17) a. What did John begin?
b. John began the reading of the book.
c. John began to read the book.
d. What did John begin to do?

The following examples show that it is possible to question the infinitival complements of verbs such as *wish*.

(18) a. What does John wish?
b. John wishes to read this book.

This behaviour of aspectual verbs resembles that of true auxiliary verbs which also cannot be solely used to question the main predicate of a sentence. Here again, in order to question the predicate, it is necessary to have recourse to the verb *do*.

(19) a. *What has John?
b. What has John done?
c. John has read a new book.

A similar behaviour of the non-finite complements of aspectual verbs is also demonstrated by the impossibility of replacing them by an anaphoric pronoun in a language like French. The only possible reference for an accusative pronoun used with *commencer* 'begin' is reference to a direct object noun phrase. The examples in (20) involving right dislocation structures illustrate these facts.

(20) a. *Jean l'a commencé, à lire le livre.
'(Lit.:) John it began, to read the book.'

b. Jean l'a commencé, son nouveau livre / la lecture du livre.
'(Lit.:) John it began, his new book / the reading of the book.'

This impossibility to refer to the infinitival complement of *commencer* by means of a pronoun cannot be attributed to the infinitival nature of the complement nor to the fact that it is introduced by the preposition *à* since other occurrences of infinitival complements introduced by this preposition can either be replaced by *le* or by *y*, depending on the grammatical function of the complement.

(21) a. Jean le lui avait enseigné, à jouer de la guitare.
'(Lit.:) John it to him had taught, to play the guitar.'

b. Jean y avait souvent pensé, à partir à l'étranger.
'(Lit.:) John *y* (to that) had often thought, to leave for a foreign country.'

Aspectual verbs also resemble auxiliary verbs in that they allow complex predicate formation (i.e., the so-called restructuring phenomenon) in languages like Italian, as shown by the following example taken from Rizzi (1978).

(22) Gianni gli continua a raccontare stupide storie.
'Gianni is continuing to tell him stupid stories.'

Lack of space does not allow me to discuss the numerous analyses that have been proposed to account for this construction. Nevertheless, I would like to point out that any explanatory analysis of this phenomenon will have to account for the fact that the verbs that can enter into this construction are verbs that do not constitute autonomous events.[8] Moreover, the different manifestations of this complex predicate phenomenon such as clitic-climbing, long object preposing and the change of auxiliary are manifestations that normally occur in monoclausal structures.

Burzio (1986) also points out that aspectual verbs like *cominciare* and *continuare* show an exceptional behaviour in that they can appear passivized with their surface subject originating as the object of the embedded verb, as shown by the following example taken from his study.

(23) Il palazzo fu cominciato a costruire sotto Carlo V.
'(Lit.:) The palace was begun to build under Charles V.'

Similar facts involving the French aspectual verbs *commencer*, *achever* and *finir* are also reported in Grevisse (1975). Whatever the correct analysis of these facts will turn out to be, it is clear that it will have to rely on a monoclausal analysis of these structures in order to account for the NP-movement of the object.

To summarize, the facts presented in this section indicate that the structures in which aspectual verbs introduce non-finite complements constitute single propositions at the level of semantic representation and monoclausal structures at the level of syntactic representation. Aspectual verbs thus appear to behave like raising predicates on a par with auxiliary verbs. I conclude that the non-finite complements introduced by aspectual verbs are VP complements. In the following section, we will see that aspectual verbs impose selectional restrictions on the event-type of their complement. The existence of such selectional restrictions also questions the validity of a bi-clausal analysis since it would then be necessary to assume that aspectual verbs can impose selectional restrictions on a portion of the "internal structure" of their complement.

The proposal that the non-finite complements of aspectual verbs should be analyzed as VP complements must however be further qualified. In fact, these non-finite complements may contain certain elements that would indicate the presence of functional projections such as NegP and AuxP over the VP. It would thus seem more appropriate to consider that the realization of the semantic category selected by the aspectual verbs is a projection of the category V, hence not precluding the possibility that these projections be embedded under some functional nodes.

This analysis also raises the question of the status of prepositions like *to* in English and *à* and *de* in French (or Italian). The conclusion reached here concerning the raising status of aspectual verbs and the VP structure of their complements clearly goes against Kayne's (1981) analysis of the French prepositions as complementizers. In Rochette (1988), I suggested that the presence of prepositions introducing the infinitival complements of aspectual verbs be linked to the fact that these verbs are Case-assigning verbs. The prepositions would be required to appear when the complement is a verbal projection rather than a nominal one. The resulting structure is however not a PP structure as shown by the fact that the infinitival complement cannot be replaced by pronouns like *y* or *en*. I rather proposed that the prepositions *à* and *de* are heading a KP (Case phrase) projection. Whatever the correct analysis of these prepositions will turn out to be, it is clear that it will have to be compatible with a raising analysis of aspectual verbs. The proposal that the complements of aspectual verbs are VP complements also goes against the current assumption that *to* is generated under I. If the presence of I (or of other elements corresponding to I) is to be equated with the presence of an autonomous event, then this analysis of *to* will have to be abandoned. I will not pursue this matter here.

3. THE SELECTIONAL RESTRICTIONS IMPOSED BY ASPECTUAL VERBS

It is a well-known fact that certain matrix verbs impose selectional restrictions on the event-types of the predicates that may appear within their non-finite complements. In particular, it has often been noted that aspectual verbs do not combine felicitously with stative verbs, as shown by the following examples.

(24) a. *John begins to own a house.
b. John begins to build a house.

The existence of such selectional restrictions raises the question of the exact nature of these restrictions and of the representation of this phenomenon in the grammar. In this section, it will be shown that aspectual verbs may freely introduce accomplishments and activities but that they generally do not allow achievements or states as their complements. Adopting the model of event semantics developed by Pustejovsky (1988;

1991), it will be proposed that activities and accomplishments form a natural aspectual class which can be characterized in terms of the semantic category *process*. It will also be proposed that the representation of the selectional restrictions imposed by aspectual verbs is best handled through the selection for this particular semantic category.

3.1. Event-types of the embedded predicates

With respect to the classification established by Vendler, it appears that predicates denoting accomplishments, such as *writing the letter*, or activities, such as *dancing*, are perfectly acceptable with all aspectual verbs, as shown by the following examples.[9]

(25) a. John began writing the letter.
b. John continued writing the letter.
c. John stopped writing the letter.
d. John finished writing the letter.

(26) a. John began dancing.
b. John continued dancing.
c. John stopped dancing.
d. John finished dancing.[10]

In contrast, predicates denoting achievements or states are generally not allowed as complements of aspectual verbs. Achievements such as *finding one's coat* or states such as *owning a house* give fairly bad results when introduced by aspectual verbs.

(27) a. *I began finding my coat.
b. *I stopped finding my coat.
c. *I finished finding my coat.

(28) a. *I began owning a house.
b. *I stopped owning a house.
c. *I finished owning a house.

This last generalization must however be further qualified since there appears to be a certain number of counter-examples noted by Freed (1979) and more particularly by Lamiroy (1987). In her study of French aspectual verbs, Lamiroy identifies five strategies that legitimate the occurrence of stative complements with aspectual verbs. These are the use of plural noun phrases, of generic noun phrases, of the "*se*-moyen", of adverbial complements and of the *par*-construction which can occur with *commencer* 'begin' and *finir* 'finish'. In what follows, I will only examine the use of plural noun phrases and that of temporal adverbial complements.

The following examples, taken from Lamiroy's study, show that the use of collective or plural noun phrases as surface subject of the aspectual verb or object of the embedded infinitive allows for occurrences of stative or achievement predicates which would not be acceptable otherwise.[11]

(29) a. *John begins to own a car.
b. John begins to own {many cars / a considerable number of cars}.

(30) a. *The guest begins arriving.
b. The guests begin arriving.

The occurrence of temporal adverbial clauses may also have the effect of allowing the use of stative or achievement predicates. The example in (31a), also taken from Lamiroy's study, and that given in (31b) are hardly acceptable without the presence of the temporal adverbial clause.

(31) a. John began to know that Ann was unfaithful to him a short time before he defended his thesis.
b. John began to arrive late the day he understood that it did not matter.

The explanation put forth by Lamiroy to account for these occurrences relies on the notion of multiplexing. Multiplexing occurs when a situation is repeated through the use of plural noun phrases or when it receives a boundary, normally absent in the case of states or achievements, through the presence of an adverbial clause allowing for a sequential interpretation of successive situations.[12]

Hence, in (29b), the presence of a plural noun phrase as object of the verb *own* has the effect of multiplying the situation described by the stative predicative complex and as such of creating a serialization which seems to put this complex on a par with activity predicates. Similarly, in (30b), the presence of a plural noun phrase as the (raised) subject of the aspectual verb also produces the same effect. In the examples in (31), the verb *begin* does not denote the initiation of a process that would lead the subject to *know* or *arrive late* but rather it denotes that the situation of *knowing* or *arriving late* has begun at the time indicated by the temporal adverbial clause. Moreover, in the case of example (31b), the event *arrive late* must necessarily receive an iterative interpretation. A further remark must be made regarding the restriction against states. This restriction appears to be sensitive to the distinction established between stage-level predicates and individual-level predicates (Kratzer (1995), Carlson (1977)). A predicate like *own*, which was used here to illustrate the restriction, is an individual-level predicate. As such, it denotes an unbounded situation and becomes acceptable with aspectual verbs only through the use of one of the strategies pointed out by Lamiroy. As expected, the presence of a temporal adverbial has the effect of allowing the stative verb to appear as a complement of aspectual verbs. Compare sentence (28a) with the following.

(32) I began owning a house the day I inherited enough money to buy one.

Stage-level predicates differ from individual-level predicates in that they can describe a bounded situation without recourse to explicit temporal adverbials to create a boundary. This is so because they describe non-permanent situations. Predicates like *love* or *sick* are normally taken to be stage-level predicates. Interestingly, they appear to combine more easily with aspectual verbs, as shown by the following examples.

(33) a. John begins to love Mary.
b. John begins to be sick.

Although these facts certainly require some further study, it seems to me that they reveal that stage-level predicates can exhibit the effect of multiplexing pointed out by Lamiroy through their non permanent characterization. As such, they would resemble activities and should probably be treated on a par with them.

It is worth mentioning at this point of the discussion that the existence of examples such as those given above clearly shows that the aspectual characterization of predicates cannot be determined without taking into account the contribution to the event structure of the other elements that constitute the predicative complex. In particular, the nature of the arguments of the predicates appears to play an important role in this matter.

A striking example of such contribution on the part of the arguments is provided to us by the contrast between examples involving the verb *find*. As was noticed above, the predicative complex *find one's coat* is not acceptable with aspectual verbs. However, when *find* occurs with a direct object such as *the solution*, it is clearly more acceptable, at least in the case of the aspectual verb *begin*.

(34) a. John began to find the solution / the answer.
b. *John stopped finding the solution / the answer.
c. ?John finished finding the solution / the answer.

The difference in acceptability between examples (27a) and (34a) is explained by the fact that *find the solution* contains a notion of process. This notion of process allows reference to the initiation and to a certain extent to the termination of the process but reference to the cessation is impossible since the achievement character of this predicative complex necessarily implies the presence of a resulting state which would not be achieved by the cessation of the process.

Freed (1979) also notes the following examples where the achievement *reach the summit* is acceptable with the verb *begin* and to a certain extent with the verb *finish*, but not with the verb *stop*.

(35) a. The American team began reaching the summit this morning.
b. *The American team stopped reaching the summit this morning.
c. ?*The American team finished reaching the summit this morning.[13]

Two factors appear to be intervening here to favour the acceptability of example (35a). First, the fact that the subject of the proposition is semantically plural has the effect of multiplying the situation described by the predicative complex, thus allowing an interpretation under which only a subset of the members of the team has reached the summit. Second, the fact that the reaching of a summit can be represented as involving a certain space or distance within which the mountain climber is in the phase of the reaching without necessarily being at the top allows us to identify a notion of process which is normally absent in the case of punctual achievements. This possibility is reminiscent of what I mentioned above with respect to the examples in (34) and also accounts for the acceptability, under this interpretation, of an example such as *John began reaching the summit this morning*.

In the preceding discussion, I purposely did not include examples of achievements occurring with the aspectual verb *continue*. This verb has the property of being associated with two distinct interpretations depending on whether it denotes the continuation or the iteration of an event.[14] Achievements are relatively acceptable with this verb but only under its iterative interpretation, which gives the achievement the character of an activity due to the repetition of the event. In this way, the following example is acceptable under the iterative interpretation of *continue* but it is not under the continuation interpretation of a single event.

(36) He continued losing his glasses.

The conclusion that can be drawn from the facts reported in this section is that aspectual verbs share the property that their complement must involve a notion of process. The notion of process is naturally present with activities and accomplishments but can only be found under certain conditions with states and achievements. Thus, in general, aspectual verbs do not accept complements denoting states or achievements.

3.2. Event Semantics

Activities and accomplishments appear to form a natural class with respect to the selection properties of aspectual verbs. I would now like to examine how the model of event semantics developed by Pustejovsky (1988; 1991) enables us to encode this fact in a non-stipulative manner.

Pustejovsky proposes that there are three primitive event-types - *states*, *processes* and *transitions* - specified in the grammar.

(37) a. State (S): *know* $[_S\ e\ ]_S$

b. Process (P): *run* $[_P\ e_1 \quad e_n\ ]_P$

c. Transition (T): *give* $[_T\ e_1 \quad e_2\ ]_T$

In this framework, *states* are characterized as single events, *processes* as repetitions of the same event and *transitions* as changes from one event-type to another. *Processes* correspond to Vendler's activities. *Transitions* can be of several sub-types depending on the event-type of the first or the second member of the transition. Achievements are analyzed as *culminating transitions* whereas accomplishments are analyzed as *logical transitions*. The main difference between these two types is that the sub-event *process* is the head (marked with a *) of a logical transition whereas in a culminating transition, the *process* is not specified and does not constitute the head of the event which is rather the second member, namely the resulting state.

(38) a. Culminating Transition: *arrive* $[_T\ [_P\ e_1 \ldots e_n\ ]_P\ \ e^*\ ]_T$

b. Logical Transition: *build* $[_T\ [_{P*}\ e_1 \ldots e_n\ ]_{P*}\ \ e\ ]_T$

Within this model, a natural class embodying activities and accomplishments can thus be defined as the set of events whose head constitutes a *process*. I propose to analyze the repetition of a state or an achievement obtained through the use of plural noun phrases or temporal adverbials as giving rise to an event structure similar to that of *processes* since these are defined as the repetition of the same event. I would also like to propose that achievements such as *find the solution* or *reach the summit*, which include an identifiable process, constitute a hybrid sub-type with respect to the opposition between culminating transitions and logical transitions. This new transition type would allow reference to the sub-type *process* to the extent that the resulting state can still be reached or is not denied by the type of reference made to the *process*.

Given this analysis, it is possible to claim that aspectual verbs are characterized as verbs that semantically select a *process*.[15] The mechanism of semantic selection enables us to encode the selectional restrictions imposed by aspectual verbs "positively" through the selection for the event-types permitted rather than "negatively" through the use of features restricting the possible realizations of a syntactic category. The semantic selection of the category *process* also makes the prediction that the same selectional restrictions should be operating across different syntactic realizations of this semantic category. As will be shown in the following section, this prediction is borne out in the case of aspectual verbs.

4. TRANSITIVE *BEGIN* AS A RAISING PREDICATE

I would now like to turn to the so-called transitive occurrences of aspectual verbs mentioned at the beginning of this paper and in section 2.1 above. I repeat the example given in (2).

(39) John began the letter.

In Rochette (1988), I argued that the existence of examples such as (39) provides further evidence in favour of the mechanism of semantic selection put forth by Grimshaw (1979) and I proposed that the canonical structural realization of the category *action* selected by effective verbs can also be a projection of N when the verb is a Case assigning verb, as is the case with aspectual verbs.

One motivation originally given by Grimshaw for the concept of semantic selection comes from the existence of examples exhibiting concealed questions or exclamations. Following observations first made by Baker (1968), Grimshaw points out that a subset of the main verbs that semantically select the category *question* not only introduce sentential indirect questions as shown in (40a) but also direct object noun phrases as shown in (40b). Interestingly, the interpretation of the direct object in (40b) is similar to that of the indirect question in (40a). Grimshaw concludes that predicates like *ask* must be associated with the semantic frame *Question* in order for the subcategorized NP to be correctly interpreted as denoting a question.

(40) a. John asked what height the building was.
b. John asked the height of the building.

Returning to example (39), it appears that its interpretation also indicates the presence of a "concealed" argument, namely a concealed *process*. Indeed, the direct object noun phrase in (39) is necessarily interpreted as the object of an implicit verb such as *write* or *read*. Interestingly, the interpretation of the nominal complements introduced by aspectual verbs can be shown to obey the selectional restrictions against states and achievement predicates that were established in the previous section. For instance, example (41a) below can receive different interpretations, like those given in (41b), but crucially these interpretations must be compatible with the restrictions imposed by *begin*, as shown by the impossible interpretations given in (41c).[16]

(41) a. John begins a new book.
b. John begins to write / read / illustrate / bind .... a new book.
c. *John begins to own / find / win .... a new book.

Obviously, the surface subject in examples (39) and (41a) is also interpreted as the subject of the implicit process predicate involved in these structures. This would therefore indicate that the surface subject of the aspectual verb is semantically the deep subject of the implicit process predicate and that these structures should also be treated as instances of raising structures. The fact that the surface subject of the aspectual verb is the subject of the implicit process predicate is further confirmed by the following sentences involving the process nominal *reading*.

(42) a. John began the reading of the book.
b. *John began the reading of the book by Mary.

The ungrammaticality of sentence (42b) follows from the fact that there can only be one subject of the process *the reading of the book*. Interestingly, aspectual verbs allow for the realization of their *process* argument in subject position when this argument is a process nominal. In the following example, the subject of the process nominal is realized through the *by*-phrase and the sentence is perfectly acceptable since there is only one subject present in contrast with (42b).

(43) The reading of the book by John has begun.

It is not obvious however how we can reconcile the fact that the two NPs in sentences such as (41a) appear to receive their thematic role from the implicit predicate and the fact that they behave as the subject and the direct object of the aspectual verb, as shown by the fact that passivization can apply to these structures.

(44) A new book was begun by John.

The interpretation of this passive sentence however still implies the presence of an implicit process predicate and the NP *John* is understood as the agent of that process. The syntactic analysis I would like to propose to capture these facts is to generate both NPs in a small clause object at D-structure, as shown in (45).[17]

(45)

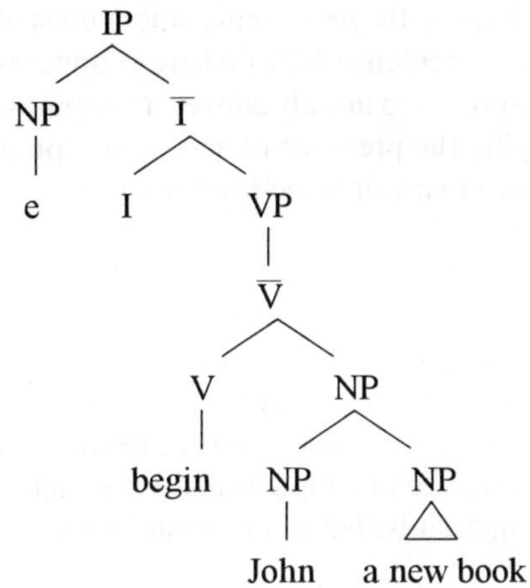

In this structure, the topmost NP that appears as a sister of the verb *begin* is the syntactic realization of the semantic category *process*. The NP *a new book* constitutes the predicative part of this small clause and will therefore denote a process compatible with the interpretation of the concrete NP *a new book* as its object. The NP *John* is the subject of the process denoted by the "predicative" NP.

Although this structure can account for the interpretation associated with the sentence, it however raises questions with respect to the attribution of the thematic roles to the two NPs and with respect to their surface grammatical functions. As for the first issue, it seems clear that the thematic roles assigned to these NPs are semantically linked to the implicit process. Whether this fact should entail that there be a syntactic representation of the implicit predicate remains an open question which will not be pursued here. As for the second issue, it appears that raising of the subject NP of the small clause to the subject position of the sentence is obligatory in the case of sentences involving "concrete objects" such as *a new book*. This would thus indicate that the accusative Case assigned by the aspectual verb is absorbed by the NP *a new book* and that the NP *John* must move to the subject position of the sentence in order to receive Case. This analysis is further confirmed by the fact that the accusative NP shows up in the subject position under passivization.[18]

The analysis presented in this section presents the advantage that it can account for the interpretation of apparent simple transitive structures which involve the presence of some implicit predicate. The cost of this analysis is that what looked like a simple transitive structure must now be analyzed as a small clause object with raising of the subject of the small clause into the subject position of the sentence. But this is not such an unexpected result given the raising analysis that appears to be necessary in the case of the non-finite complements.

There remain several questions with respect to the nominal realization of the *process* argument of aspectual verbs. For instance, the example given in (43) shows an intransitive use of aspectual verbs. In such cases, the *process* argument must obligatorily appear in subject position in contrast with sentence (42a) where it appears in object position. Also the aspectual verb in (43) appears to mainly convey a temporal indication with respect to its subject perhaps implying the presence of an implicit predicate such as *to take place*. Such questions will have to await some further study.

5. CONCLUSION

The facts discussed in the preceding sections all point to the mixed behaviour of aspectual verbs when compared with main verbs and with auxiliary verbs. On the one hand, aspectual verbs, like other matrix verbs, introduce non-finite complements and can appear as the unique predicate of a sentence in simple transitive structures. On the other hand, like auxiliary verbs, their function is that of predicate modifiers and they do not select a subject.

The raising analysis of aspectual verbs together with the proposal that these verbs select a *process* which is realized either as a projection of V or a projection of N can account for the mixed behaviour of these verbs. Under this analysis, aspectual and auxiliary verbs share the property of appearing in monoclausal structures. Aspectual verbs however are treated as predicates selecting an internal argument and in that sense they resemble other matrix verbs although the syntax of the structures in which they appear is certainly not as transparent as it first seemed to be.

NOTES

[*] This paper, like all the papers in this volume, is dedicated to the memory of Osvaldo Jaeggli. My interest in the topic of aspectual verbs originally grew out of numerous conversations with Osvaldo when we were students at MIT. I would like to take this opportunity to express my indebtedness for his encouragement and friendship.

This research has been supported by grants from the Social Sciences and Humanities Research Council of Canada (Research Grant 410-92-1434), the *Fonds pour la formation de chercheurs et l'aide à la recherche* (Research Grant 93-NC-0978) and the Université du Québec à Montréal.

[1] See, among others, Perlmutter (1970) and Ruwet (1972). The validity of positing two subcategorization frames for aspectual verbs has been questioned by Newmeyer (1975) who argues in favour of a raising analysis of these verbs. See also Ruwet (1983) with respect to the distinction raising/control.

[2] Lack of space prevents me from reviewing the arguments put forth by Grimshaw, but see section 4 below for the argument concerning the existence of "concealed questions".

[3] Following Pesetsky (1982), I also assumed that categorial selection can be derived from semantic selection. It may be that this assumption will be impossible to maintain across the grammar. Semantic selection however will remain necessary to account for the interpretation of sentences such as (2), as will be shown in section 4.

[4] Propositional predicates sometimes introduce subjunctive complements but only in contexts of negation or direct question. The subjunctive complements introduced by emotive predicates differ from those governed by the negation or the interrogation of the propositional predicates in terms of several properties such as the coreference possibilities between the embedded and the matrix subjects (see Rochette (1988)).

[5] The different classes of predicates distinguished here should be viewed as constituting a continuum of the following type: effective < emotive < propositional. Under such a view, the predicates inside each class could also be organized in a continuum and aspectual verbs would be situated at the leftmost limit of the effective class or perhaps between true auxiliary verbs and effective verbs.

[6] Similar examples are given by Rouveret and Vergnaud (1980) with respect to the verb *menacer* 'threaten'. Interestingly, this verb also imposes selectional restrictions on its non-finite complement and does not select a proposition under its raising interpretation.

[7] See also Newmeyer (1975) who points out similar facts.

[8] See Rochette (1990) for an analysis of this construction that takes into account this property of the so-called restructuring verbs.

[9] Aspectual verbs in English show the interesting property that some of them allow both infinitive and gerundive complements while others only allow for one type. I will not study this phenomenon in this paper. See Freed (1979) on this issue.

[10] It is not clear that this example is truly acceptable under a strict activity interpretation of *dancing*. In general, the aspectual verb *finish* is rather odd with activities, as noticed by Freed (1979). It seems that the notion of completion involved with this verb requires an event-type for which there is a logical culmination at the end of the event, namely an accomplishment. With respect to the analysis developed at the end of section 3.2 below, this would mean that *finish* imposes an additional restriction on its complement; this verb would not only semantically select a *process* but would also require that there be an identifiable resulting state.

[11] The restrictions appear to be the same in English and French. I therefore take the liberty to give the English translations of Lamiroy's examples.

[12] Lack of space does not allow me to go into the details of Lamiroy's analysis but I would like to mention that her conclusion is that when the aspectual verbs occur with stative predicates as a result of one of the above mentioned operations, they do not express aspect but rather belong to the temporal subcategory of serialization.

[13] The grammaticality judgments given here are those of Freed. I personally find that example (35c) is no more marginal than example (34c) and I would assign it a simple question mark.

[14] According to Lamiroy (1987: 285), while the continuative meaning of this verb is a basic property of aspect, the iterative meaning should be "considered as a special case of the temporal subcategory which locates different events with relation to each other - i.e. serialization."

[15] Voorst (1993) develops a localist approach to event semantics in which the notions of *inclusion* and *cohesion* play a predominant role in the aspectual characterization of predicates. Within this model, activities and accomplishments may also form a natural class since they both are characterized by the fact that they involve cohesion. It may however be more difficult to pursue an approach in terms of selection of a particular semantic category, unless the notion of cohesion can be conceived of as a conceptual category constituting a semantic primitive.

[16] In fact, it seems that not only states and achievements but also activities are disallowed as possible candidates for the interpretation of the implicit predicates. For instance, Newmeyer (1975) notices that a sentence like (ia) can only receive interpretations involving an accomplishment as in (ib) and not an activity as in (ic).

(i) a. The man began the sermon.
 b. The man began delivering the sermon.
 c. The man began listening to the sermon.

I believe that this further restriction is due to the presence of the direct object NP which is interpreted at the "end point" of the process and therefore adds a telicity to the interpretation of the concealed process.

[17] The IP and NP projections are used here as a means to simplify the structure and should not be interpreted as a claim against the existence of functional projections such as AgrP, TP and DP.

[18] This proposal implies that there are verbs that assign accusative Case while not selecting an external argument, contrary to Burzio's generalization.

REFERENCES

Baker, C.L.: 1968, *Indirect Questions in English*, Doctoral Dissertation, University of Illinois, Urbana, Illinois.

Burzio, L.: 1986, *Italian Syntax*, Reidel, Dordrecht.

Carlson, G.N.: 1977, *Reference to Kinds in English*, Doctoral Dissertation, University of Massachusetts, Amherst.

Chomsky, N.: 1965, *Aspects of the Theory of Syntax*, MIT Press, Cambridge, Massachusetts.

Freed, A. F.: 1979, *The Semantics of English Aspectual Complementation*, Reidel, Dordrecht.

Grevisse, M.: 1975, *Le Bon Usage*, Editions J. Duculot, Gembloux.

Grimshaw, J.: 1979, 'Complement Selection and the Lexicon', *Linguistic Inquiry* 10, 279-426.

Kayne, R. S.: 1981, 'On Certain Differences between French and English', *Linguistic Inquiry* 12, 349-371.

Kratzer, A.: 1995, 'Stage-level and Individual-level Predicates', In G. N. Carlson and F. J. Pelletier, eds., *The Generic Book*. The University of Chicago Press, Chicago, 125-175.

Lamiroy, B.: 1987, 'The complementation of aspectual verbs in French', *Language* 63, 278-298.

Long, M. E.: 1974, *Semantic Verb Classes and their Role in French Predicate Complementation*, Doctoral Dissertation, University of Indiana, Bloomington. Distributed from 1976 by Indiana University Linguistics Club, Bloomington.

Newmeyer, F. J.: 1975, *English Aspectual Verbs*, Mouton, The Hague.

Perlmutter, D. M.: 1970, 'The Two Verbs *Begin*', in R. Jacobs and P. Rosenbaum, eds., *Readings in English Transformational Grammar*, Ginn, Waltham, Massachusetts, 107-120

Pesetsky, D.: 1982, *Paths and Categories*, Doctoral Dissertation, MIT, Cambridge, Massachusetts.

Pustejovsky, J.: 1988, 'The Geometry of Events', in C. Tenny, ed., *Studies in Generative Approaches to Aspect, Lexicon Project Working Papers* 24, MIT, Cambridge, Massachusetts, 19-39.

Pustejovsky, J.: 1991, 'The Syntax of Event Structure', *Cognition* 41, 47-81.

Rizzi, L.: 1978, 'A Restructuring Rule in Italian Syntax', in S.J. Keyser, ed., *Recent Transformational Studies in European Languages*, The MIT Press, Cambridge, Massachusetts, 113-158.

Rochette, A.: 1988, *Semantic and Syntactic Aspects of Romance Sentential Complementation*, Doctoral Dissertation, MIT, Cambridge, Massachusetts.

Rochette, A.: 1990, 'On the Restructuring Classes of Verbs in Romance', in A.-M. Di Sciullo and A. Rochette, eds., *Binding in Romance. Essays in Honour of Judith McA'Nulty*, Special Publication of The Canadian Linguistic Association, Ottawa, 96-128.

Rouveret, A. and J.-R. Vergnaud: 1980, 'Specifying Refrence to the Subject: French Causatives and Conditions on Representations', *Linguistic Inquiry* 11, 97-202.

Ruwet, N.: 1972, *Théorie syntaxique et syntaxe du français*, Le Seuil, Paris.

Ruwet, N.: 1983, 'Montée et contrôle: Une question à revoir?', *Revue Romane* 24, 17-37.

Vendler, Z.: 1967, *Linguistics and Philosophy*, Cornell University Press, Ithaca.

Voorst, J. van: 1993, 'A Localist Model for Event Semantics', *Journal of Semantics* 10, 65-111.

SUBJECT PREDICATION WITHIN IP AND DP*

Mamoru Saito and Keiko Murasugi
Nanzan University

1. INTRODUCTION

According to the VP-internal subject hypothesis, proposed by Fukui and Speas (1986), Kuroda (1988), Koopman and Sportiche (1985), and Kitagawa (1986), among others, the subject of a finite sentence is generated within VP and is raised to the IP Spec position at S-structure to have its Case checked. Fukui and Speas (1986), Abney (1986), and Kuroda (1988) propose a parallel analysis for the subject of NP: it is generated within NP and is moved to DP Spec. The S-structures of (1a-b), then, are as in (2a-b), respectively.[1]

(1) a. the barbarians destroyed the city
    b. the barbarians' destruction of the city

(2) a. $[_{IP}$ the barbarians$_1$ $[_{\bar{I}}$ I $[_{VP}$ t_1 $[_{\bar{V}}$ destroyed the city]]]]

    b. $[_{DP}$ the barbarians'$_1$ $[_{\bar{D}}$ D $[_{NP}$ t_1 $[_{\bar{N}}$ destruction of the city]]]]

This paper consists of two notes on the VP/NP-internal subject hypothesis illustrated above. In the following section, we will discuss two phenomena that suggest the possibility of the structures in (3).

(3) a. $[_{IP}$ the barbarians$_1$ $[_{\bar{I}}$ I $[_{VP}$ PRO$_1$ $[_{\bar{V}}$ destroyed the city]]]]

    b. $[_{DP}$ the barbarians'$_1$ $[_{\bar{D}}$ D $[_{NP}$ PRO$_1$ $[_{\bar{N}}$ destruction of the city]]]]

The idea that PRO can occur in NP Spec within a DP is entertained in Fukui and Speas (1986), and Giorgi and Longobardi (1991). In 2.1-2.2, we will examine the phenomenon of NP-deletion in some detail, and argue further that there can be a control relation between a phrase in DP Spec and PRO in NP Spec, as in (3b). Then, in 2.3, we will discuss VP-preposing, and suggest that the same kind of control relation is possible within a finite IP. As far as we know, the hypothesis that the subject of a sentence is predicated of the VP (or $\bar{I}$) via control was first seriously considered by David Lebeaux.[2] Thus, the discussion in this paper provides empirical support for his general idea. It further provides evidence for the proposal in Fukui and Speas (1986) that PRO, unlike lexical DPs, can stay within VP at S-structure.

K. Johnson and I. Roberts (eds.), Beyond Principles and Parameters, 167-188.

Section 3 is concerned with the absence of NP-preposing and IP-preposing, as opposed to VP-preposing. Given the parallelism between IPs and DPs, the absence of NP-preposing seems particularly surprising. Or put differently, if IPs and DPs indeed have parallel structures as in (3) (or as in (2)), we would expect the absence of NP-preposing to follow from independently motivated principles. We will argue that the expectation is fulfilled. We will first show that all cases of IP-preposing and most cases of NP-preposing are ruled out by the Head Licensing Condition on empty categories, originally proposed by Jaeggli (1982), along the lines suggested by Zagona (1982) and Lobeck (1990). Then, we will argue that the remaining cases of NP-preposing are excluded by the "antecedent government requirement," or the ECP as formulated in Chomsky (1991) and Chomsky and Lasnik (1993). The discussion in this section provides support for those principles, as well as for the parallel analysis of IPs and DPs.

2. NP/VP-INTERNAL PRO

In this section, we will argue for the structures in (3), based on the phenomena of $\overline{N}$-deletion and VP-preposing. But first, we will briefly go over the arguments for the reanalysis of $\overline{N}$-deletion as NP-deletion in Section 2.1.

2.1. $\overline{N}$-Deletion and the DP Hypothesis

The $\overline{N}$-deletion phenomenon is studied extensively in Jackendoff (1971). A typical example is shown in (4).

(4) I like Bill's wine, but Max's is even better.

Lobeck (1990), and Saito and Murasugi (1990) independently argue that this phenomenon provides support for the DP hypothesis mentioned above. Lobeck (1990), in particular, discusses the parallelism among $\overline{N}$-deletion, VP-deletion and Sluicing in detail, and shows that the DP hypothesis enables us to maintain the following generalizations:

(5) a. The three phenomena involve "deletion" of the complement of a functional head.
b. The "deletion" is licensed only by a functional head that agrees with its Spec.

The examples in (6)-(8) illustrate these generalizations. (The contrast in (7) is due to Takahashi (1994) and Martin (1992).)

(6) a I read John's book, and now, I want to read Mary's.
b. *I read about that person, and now, I want to see the.

(7) a. I believe that Mary is smart, and I believe that John is, too.
b. *I believe Mary to be smart, and I believe John to, too.

(8) a. I know that Mary bought something, but I don't know what.
b. *Mary said that she was going to Boston, but I don't know whether.

(7a) and (8a) involve "deletion" of the complements of I and C, respectively. And given the DP hypothesis, the structure of (6a) can be as in (9).

(9) ... $[_{DP}$ Mary's $[_{\overline{D}}$ D $[_{NP}$ $\underline{e}]]]$

Thus, what is "deleted" in (6a) is the complement of a functional head D. Under this hypothesis, $\overline{N}$-deletion is reanalyzed as NP-deletion. (7b)-(8b) illustrate the generalization in (5b). (7b) shows that VP-deletion is allowed only when the I agrees with its Spec.[3] (8b) indicates that Sluicing has the same general property: an IP can be "deleted" only when the C agrees with its Spec. If we assume the standard $\overline{N}$-deletion analysis, the contrast in (6) simply suggests that $\overline{N}$-deletion is allowed only when there is a genitive phrase. But if we assume the DP hypothesis/NP-deletion analysis, the contrast readily falls under the generalization in (5b). Since *the* is not an agreeing functional head, it fails to license NP-deletion.

In Saito and Murasugi (1990), we presented two pieces of empirical evidence for the DP hypothesis/NP-deletion analysis, in addition to the conceptual argument illustrated above. The first piece of evidence has to do with Anderson's (1983) paradigm, to which we will return in the following subsection. The other piece of evidence has to do with $\overline{N}$(NP)-deletion in Japanese.

In this language, the genitive Case marker *no* is not assigned to any specific position, but appears on any NP(DP) or PP that is immediately dominated by a projection of N (or D). Thus, the so called multiple genitive construction is possible as shown in (10), and adjunct NP(DP)s are also accompanied by the genitive Case marker as shown in (11).

(10) a. yuubokumin-no tosi-no hakai
nomad -gen city-gen destruction
'the nomad's destruction of the city'

b. John-no Mary-e -no izon
-gen -on -gen reliance
'John's reliance on Mary'

(11) a. ame -no hi
rain -gen day
'rainy day'

b. hutakire -no hamu
two slice -gen ham
'two slices of ham'

Hence, on the surface, there is no clear difference between arguments and adjuncts within NPs.

However, interestingly enough, only argument genitive phrases license $\overline{N}$(NP)-deletion. For example, (12a-b) are well-formed, but (13a-b) are not.

(12) a. [rooma-no hakai] -wa [kyooto-no e̲] yorimo hisandatta.
Rome -gen destruction-top Kyoto -gen than horrifying-was
'Rome's destruction was more horrifying than Kyoto's.'

b. [gakubusei -no sensei -e -no izon]-wa yuruseru
undergraduate-gen teacher-on-gen dependence-top tolerable-is

ga, [insei -no e̲]-wa yurusenai.
though graduate-gen -top tolerable-not-is
'The undergraduates' dependence on the faculty is tolerable, but the graduate students' is not.'

(13) a. *saikin -wa [hare -no hi]-ga [ame -no e̲] yorimo ooi.
recently-top clear-gen day-nom rain-gen than plentiful-is
'Recently, there have been more clear days than rainy days.'

b. *[hutakire -no hamu]-wa yuusyoku-ni naru ga,
two slice-gen ham -top supper -to make though

[hitokire -no e̲]-wa naranai.
one slice-gen -top make-not
'Two slices of ham make up a supper, but one slice does not.'

If we assume the standard $\overline{N}$-deletion analysis and the simple generalization that a genitive phrase licenses $\overline{N}$-deletion, this contrast seems quite mysterious. However, given the DP hypothesis/NP-deletion analysis, this contrast can be related to the general inability of adjuncts to appear in the DP Spec position. As illustrated in (14-15), adjuncts in general cannot move from an NP-internal position to DP Spec.[4]

(14) a. the books there
b. *there's books

(15) a. the brown book
b. *brown's book

Hence, *ame-no* in (13a), for example, cannot move to DP Spec, and consequently, cannot agree with D to license NP-deletion. On the other hand, *kyooto-no* in (12a), being an argument, can move to DP Spec, and participate in Spec/head agreement with D. Thus, given the DP hypothesis/NP-deletion analysis, the contrast between (12) and (13) is exactly what we expect. This contrast then provides strong evidence for the existence of DP in Japanese, and for the DP hypothesis in general. In the following subsection, we will present evidence for the structure in (3b), based on the NP-deletion phenomenon.

2.2. Anderson's (1983) Paradigm

Anderson (1983) discusses the following paradigm in relation to the possible distribution of empty nominals:

(16) a. This book is John's e.
b. *That reliance on friends is Mary's e.
c. *That destruction of the city is the barbarians' e .

We argued in Saito and Murasugi (1990) that this contrast constitutes further supporting evidence for the DP hypothesis/NP-deletion analysis. The "predeletion" structures of (16a-b), for example, are as in (17a-b) respectively.

(17) a. $[_{DP}$ This $[_{NP}$ book]] is $[_{DP}$ John's $[_{NP}$ book]].

b. $[_{DP}$ This $[_{NP}$ reliance on friends]] is $[_{DP}$ Mary's$_1$ $[_{NP}$ t_1 reliance on friends]].

The second NP is identical to the first in (17a), but not in (17b). Hence if NP-deletion requires a syntactic antecedent, the contrast between (17a) and (17b) directly follows.

However, as also noted in Saito and Murasugi (1990), there are examples that are potentially problematic for this analysis. For example, (18) and its English counterpart (19) are both fine.

(18) [[John-ga Mary-ni yoseru] sinrai]-wa [Bill-no e] yorimo atui.
-nom -to have trust -top -gen than deep

(19) The trust that John has in Mary is deeper than Bill's e.

In this subsection, we will argue that examples like (18)-(19) constitute evidence for the structure in (3b).

Suppose first that the "pre-deletion" structure of (19) is as in (20).[5]

(20) [$_{DP}$ The [$_{NP}$ [$_{NP}$ PRO$_1$ trust] [that John$_1$ has in Mary]]] is deeper than
[$_{DP}$ Bill's$_1$ [$_{NP}$ t_1 trust]].

Then, it may seem possible that the deletion in (19) is allowed because *[$_{NP}$ PRO trust]* and *[$_{NP}$ t trust]* count as "identical" despite the difference between PRO and trace. But this analysis is untenable: "deletion" seems sensitive to the distinction between PRO and trace, as we will see directly.

The following example from Lasnik and Saito (1992) shows that *promise* allows not only control, but also raising:

(21) ?Headway$_1$ promises [t_1 to be made by John].

Thus, (22a-b) are also ambiguous between the control and the raising readings.

(22) a. John$_1$ promises [$\underline{e}_1$ to be successful].

b. John$_1$ [$_{VP}$ promises [$\underline{e}_1$ to be successful]], and Mary does [$_{VP}$ e], too.

But in (22b), the second conjunct must have the same reading as the first. That is, the first and the second conjuncts must receive identical interpretation with respect to control vs. raising. This fact indicates that PRO and trace do not count as "identical" with respect to deletion phenomenon. The following examples illustrate this point more clearly.

(23) a. John$_1$ promises [t_1 to be successful], and Mary$_1$ actually promised
[PRO$_1$ to be successful].
b. *John promises [t_1 to be successful], and Mary actually did.

(24) a. John$_1$ often promises [PRO$_1$ to be successful], but Mary is the
one who$_1$ really promises [t_1 to be successful].
b. *John$_1$ often promises [PRO$_1$ to be successful], but Mary is the
one who really does.

(23b)-(24b) are out as the VP-deletion versions of (23a)-(24a) since the "deleted" VPs do not have syntactic antecedents. (20), thus, cannot be the "pre-deletion" structure of (19).

Then, what can be the "pre-deletion" structure of (19)? Since the subject of the relative head *trust* cannot be a trace and must be PRO, the only possibility seems to be the following:

(25) [$_{DP}$ The [$_{NP}$ [$_{NP}$ PRO$_1$ trust] [that John$_1$ has in Mary]]] is deeper than
[$_{DP}$ Bill's$_1$ [$_{NP}$ PRO$_1$ trust]].

This structure implies that the subject of a DP can be base-generated in DP Spec, and be coindexed with PRO in the NP-internal subject position, as illustrated in (26).

(26) $[_{NP}$ DP$_1$ $[_{NP}$ PRO$_1$ $[_{\bar{N}}$...]]]

This analysis was in fact suggested in Saito and Murasugi (1990), but was not ultimately adopted as it was inconsistent with our analysis of Anderson's (1983) paradigm in (16). Once we admit the structure in (26), nothing seems to prevent (16b), for example, from having the "pre-deletion" structure in (27).

(27) $[_{DP}$ This $[_{NP}$ PRO reliance on friends]] is $[_{DP}$ Mary's$_1$ $[_{NP}$ PRO$_1$ reliance on friends]].

Thus, we lose the NP-deletion account for the examples in (16b-c). These examples, then, become problematic if we accept the structure in (26).

In the remainder of this subsection, we will argue that (19) in fact can have the "pre-deletion" structure in (25), and that the structure in (26) is indeed possible. We will suggest that examples like (16b-c), contrary to Saito and Murasugi (1990), are to be ruled out independently of NP-deletion.

First, as Ray Jackendoff (p.c., 1989) has pointed out to us, the range of examples that seem to require the structure in (26) is quite extensive.[6] A few more examples are shown in (28)-(29).

(28) a. [This attack on Rome] took longer than [the enemy's e].
b. [That attack on Rome] was successful, but [the enemy's e] wasn't.

(29) [song kenkyuu -ni taisuru taido] -wa [Taroo-no e] yorimo warui.
that research -toward attitude-top -gen than bad
'That attitude toward research is worse than Taroo's.'

Given that "deletion" is sensitive to the distinction between trace and PRO, the "pre-deletion" structure of (28a), for example, must be as in (30).

(30) $[_{DP}$ This $[_{NP}$ PRO attack on Rome$_1$] took longer than $[_{DP}$ the enemy's$_1$ $[_{NP}$ PRO$_1$ attack on Rome]].

Hence, these examples, like (18)-(19), provide evidence for the structure in (26).

Secondly, (16b-c) are bad even without NP-deletion, as illustrated in (31).[7]

(31) a. This book is John's book.
b. *That reliance on friends is Mary's reliance on friends.
c. *That destruction of the city is the barbarians' destruction of the city.

This fact clearly indicates that (16b-c) should be ruled out independently of whether or not the structure in (26) is possible. This is so, since the ungrammaticality of (31b-c) shows that (16b-c) are ruled out regardless of whether the deletion is legitimate with a linguistic antecedent for the "deleted" NP. Hence, those examples, after all, do not pose a problem for the structure in (26).

We speculate here that the ill-formedness of (31b-c) is related to a specific property of equative sentences. If a Θ-role assigning noun, like *reliance* and *destruction*, must have a subject, then the structure of (31b), for example, will be as in (27), repeated below as (32).

(32) $[_{DP}$This $[_{NP}$PRO reliance on friends]] is $[_{DP}$Mary's$_1$ $[_{NP}$PRO$_1$ reliance on friends]].

Since (31b) is an identity statement, the PRO in the first DP in (32) must be coindexed with, and referentially dependent on, *Mary* in the second DP. But it seems that the reference of the first DP in an equative sentence must be established independently of the rest of the sentence. Thus, it seems to be generally the case that any element in the first DP cannot be referentially dependent on any element in the second DP. The following examples illustrate this fact:

(33) a. *His$_1$ book is John's$_1$ book.
b. *Her$_1$ reliance on friends is Mary's$_1$ reliance on friends.

(34) *[The book that he$_1$ was reading] is John's$_1$ book.

Hence, given the structure in (32), (31b) will be ruled out for the same reason as (33)-(34).

As should be clear from the discussion above, Anderson's (1983) paradigm in (16) seems neutral with respect to the structure in (26). We conclude, therefore, that (25) can be the "pre-deletion" structure of (19), and more generally, that there can be a control relation between a phrase in DP Spec and PRO in NP Spec.

2.3. Controlled PRO in the Spec of VP

In the preceding subsection, we argued that DPs can have the following structure based on some examples of NP-deletion:

(35) $[_{DP}$DP$_1$ $[_{NP}$PRO$_1[_{\overline{N}}$...]]]

Given the parallelism between DP and IP, we expect then that IPs can have the structure in (36).

(36) $[_{IP}$DP$_1$ $[_{VP}$PRO$_1$ $[_{\overline{V}}$...]]]

In this section, we will discuss VP-preposing, and suggest that this is indeed the case.[8]

Let us first consider the following examples:

(37) a. [How likely [t_1 to win]]$_j$ is John$_1$ t_j?
b. *[Which picture of t_1]$_2$ does John wonder who$_1$ Mary bought t_2?

In Saito (1986), the contrast in (37) was taken as evidence that "reconstructions" (or chain binding in the sense of Barss (1984)) obtains with A-binding, but not with A'-binding. According to this hypothesis, the trace t_1 of raising in (37a) is "A-bound" by its antecedent due to "reconstruction," exactly as the reflexive in (38).

(38) [Which picture of himself$_1$]$_2$ does John$_1$ like t_2 best?

The trace t_1 of wh-movement in (37b), on the other hand, needs to be A'-bound. As A'-binding is insensitive to "reconstruction," this trace must have a c-commanding antecedent in the strict sense. The example is, thus, ruled out by the Proper Binding Condition.

However, Anthony Kroch (p.c., 1986) has pointed out the following examples, originally due to Mark Baltin, which indicate that traces of raising do not show reconstruction effects:

(39) a. *[How likely [t_1' to be taken t_1 of John]]$_2$ is advantage$_1$ t_1?
b. *[How likely [t_1 to be a riot]]$_2$ is there$_1$ t_2?

As the idiom chunk *advantage* and the expletive *there* occupy the matrix subject position in (39a-b), these examples, unlike (37a), clearly involve raising. The ungrammatical status of these examples, then, suggests that traces of raising, like those of wh-movement, are insensitive to "reconstruction." Given this observation, it is argued in Lasnik and Saito (1992) that the Proper Binding Condition, as opposed to Condition (A), requires strictly that traces have c-commanding antecedents, and that (37a) has the structure in (40). (See also Saito (1989).)

(40) [How likely [PRO$_1$ to win]]$_2$ is John$_1$ t_2?

Let us now consider the following examples of VP-preposing in the light of this conclusion:[9]

(41) a. John said he would win the race, and win the race, he did.
b. Mary said she would be praised by the critics, and praised by the critics, she was.

According to the standard VP-internal subject hypothesis, the structure of the second conjunct in (41a) is as in (42).

(42) $[_{VP}\, t_1$ win the race $]_2$, he_1 did t_2.

But then, the example should be ruled out by the Proper Binding Condition, since the trace t_1 is not c-commanded by its antecedent. The well-formedness of the example, thus, indicates that the empty category in the VP Spec position need not be a trace. The only option for the example, in fact, seems to be that it is PRO, exactly as the embedded subject in (40). We conclude, then, that the structure of the second conjunct in (41a) can be as in (43).

(43) $[_{VP}$ PRO_1 win the race $]_2$, he_1 did t_2.

This analysis of (41a) carries over directly to (41b). The standard VP-internal subject hypothesis assigns the following structure to the second conjunct of this example:

(44) $[_{VP}$ (t_1') praised t_1 by the critics $]_2$, she_1 was t_2.

Thus, we would falsely predict the example to be ungrammatical. But once we assume that PRO can appear in VP Spec as in (43), the structure in (45) becomes possible.

(45) $[_{VP}$ PRO_1 praised t_1 by the critics $]_2$, she_1 was t_2.

(41b) is then correctly predicted to be grammatical.

The analysis proposed above predicts that VP-preposing is impossible when the IP Spec is occupied by an idiom chunk. The analysis of (39a), repeated below as (46), implies that an idiom chunk cannot be base-generated in IP Spec and control a PRO, but must be raised to that position leaving behind a trace.

(46) *[How likely [$\underline{e}_1'$ to be taken $\underline{e}_1$ of John$]]_2$ is $advantage_1$ t_2?

Hence, when an idiom chuck is in the IP Spec, VP-preposing should result in a Proper Binding Condition violation. Here, there are some potentially problematic cases. But the prediction is borne out to a large extent, as shown by the following examples:

(47) a. *They said headway would be made by her, and made by her, it (headway) was.
b. *They said tabs would be kept on her, and kept on her, they (tabs) were.

The second conjuncts of (47a-b) have the structures in (48a-b) respectively.

(48) a. $[_{VP}$ (t_1') made t_1 by her$]_2$, it_1 (headway) was t_2.

b. $[_{VP}$ (t_1') kept t_1 on her $]_2$, $they_1$ (tabs) were t_2.

Thus, they are ruled out by the Proper Binding Condition, exactly as (49a-b).

(49) a. *[How likely to be made by Mary]$_2$ is headway t_2?
b. *[How likely to be kept on Mary]$_2$ are tabs t_2?

The examples in (47), then, provide further evidence for our analysis.[10]

If the argument above is on the right track, there must be a control relation between the phrase in IP Spec and PRO in VP Spec in all VP-preposing examples. Put differently, it is this control relation that makes VP-preposing possible. Then, more generally, IPs must be able to have the structure in (3a), repeated in (50), which parallels the structure of DP in (3b), repeated in (51).

(50) $[_{IP}$ the barbarians$_1$ $[_{\bar{I}}$ I $[_{VP}$ PRO$_1$ $[_{\bar{V}}$ destroyed the city$]]]]$

(51) $[_{DP}$ the barbarians'$_1$ $[_{\bar{D}}$ D $[_{NP}$ PRO$_1$ $[_{\bar{N}}$ destruction of the city$]]]]$

In the following section, we will consider a difference between an IP-internal VP and a DP-internal NP. Despite the parallelism between VP and NP in (50)/(51), NP-preposing, in distinction to VP-preposing, is uniformly barred. We will argue that all cases of NP-preposing are ruled out by independently motivated principles, and hence, this difference does not constitute a problem for the parallel analysis of IPs and DPs.

3. ON THE ABSENCE OF IP/NP-PREPOSING

In the preceding section, we suggested that the subject in IP Spec can control a PRO in VP Spec as in (50), and that this is the reason VP-preposing is possible without violating the Proper Binding Condition. Since we also suggested that DPs can have the structure in (51), NP-preposing out of a DP, like VP-preposing, need not violate the Proper Binding Condition. But NP-preposing is impossible as shown in (52).

(52) a. *Book$_1$, I read [John's t_1].

b. *Destruction of the city$_1$, I witnessed [the barbarians' t_1].

The main concern of this section is the absence of NP-preposing, as opposed to VP-preposing.[11]

More generally, as will be shown below, though VP, NP, and IP can all be "deleted," only VP can be preposed. We will show in this section that IP-preposing and NP-preposing are both ruled out by independently motivated principles. In the following subsection, we will argue that all cases of IP-preposing and most cases of NP-preposing are excluded by the Head Licensing Condition on empty categories. Then, in Section 3.2, we will argue that the "antecedent government requirement," or the ECP as formulated in Chomsky (1991) and Chomsky and Lasnik (1993), rules out the remaining cases. The analysis proposed in this section provides support for those principles, and also shows that the difference between VP and NP/IP with respect to the possibility of proposing is quite consistent with the parallel analysis of the functional projections, IP, DP, and CP.

3.1. The Head Licensing Requirement on Empty Categories

It is argued in Zagona (1982) that VP-deletion and VP-preposing are allowed in the same contexts. The examples in (53)-(54) illustrate this generalization.

(53) a. I believed that Mary would win the race, and she did.
b. I believed that Mary would win the race, and win the race, she did.

(54) a. *Mary believes Bill to be smart, but I don't believe him to.
b. *Mary believes Bill to be smart, but be smart, I don't believe him to.

Recall here that the VP-deletion examples fall under Lobeck's (1990) generalization in (5b), restated in (55).

(55) A functional head licenses the deletion of its complement only when it agrees with its Spec.

Given the parallelism between "deletion" and preposing, the relevant generalization should be stated more generally as in (56), as in fact suggested by Zagona (1982) for VP-deletion and VP-preposing.

(56) A functional head licenses an empty category in its complement position only when it agrees with its Spec.

Following Zagona and Lobeck, we assume here that (56) is to be explained by the so-called "head government requirement on empty categories," originally proposed by Jaeggli (1982).[12] For the purpose of discussion, we adopt the following formulation of this condition:

(57) The Head Licensing Condition on Empty Categories[13]
a. An empty category must be within a projection of a licensing head.
b. X is a licensing head iff it is lexical or agrees with its Spec.

This condition excludes (54) since the embedded Infl in the second conjunct is not a licensing head. On the other hand, the empty VPs in (53) satisfy this condition: the relevant Infl agrees with the subject, and hence, is a licensing head.

As far as we can see, the condition in (57) has the effect of excluding all cases of IP preposing. Let us first consider the following example:

(58) *$[_{IP}$ Bill will visit tomorrow $]_1$, I think that t_1 .

Chomsky and Lasnik (1993) discuss this example as evidence that Comp, when it contains *that*, is not a proper (head) governor. In our terms, the trace in (58) fails to be licensed because the Comp does not agree with its Spec and hence is not a licensing head. Thus, (58) is ruled out in the same way as the example of Sluicing in (59).

(59) *Mary thinks that Bill will visit tomorrow, but I don't think $[_{CP} [_{\overline{C}}$ that $[_{IP} \underline{e}]]]$.

Note here that (58) is out even without the complementizer *that*. This indicates that the representation in (60) is also in violation of (57).

(60) *$[_{IP}$ Bill will visit tomorrow $]_1$, I think $[_{CP}\ t_1'\ [_{\overline{C}}\ [_C \underline{e}]\ t_1]]$.

And this, in turn, implies that an intermediate trace in CP Spec does not agree (in the relevant sense) even with an empty Comp.[14] This conclusion is confirmed by the following ill-formed example of Sluicing:

(61) *Bill thinks someone stole the car, but I don't know who_1 he thinks $[_{CP}\ t_1'\ [_{\overline{C}}\ [_C \underline{e}]\ [_{IP} \underline{e}\ ]]]$.

Given the condition in (57), then, IP-preposing, like Sluicing, is impossible without an operator in CP Spec. This is so, since otherwise, the head C will fail to license the trace. On the other hand, if the operator moves to CP Spec from a position within the IP, the IP-preposing necessarily results in a Proper Binding Condition violation. The example in (62) illustrates this point.

(62) *$[_{IP}$ John bought $t_2\ ]_1$, I don't know $[_{CP}\ what_2\ [_{\overline{C}}\ [_C \underline{e}]\ t_1]]$.

The wh-phrase *what* in CP Spec turns the embedded COMP into a licensing head, and hence, the trace of IP is licensed. But once *what* moves into the CP Spec and the IP is proposed, the trace of the wh-phrase fails to be bound. Thus, the resulting configuration is ruled out by the Proper Binding Condition.

The discussion above implies that if IP-preposing is possible at all, we must have a base-generated operator in CP Spec. Here, *whether* and *if* may be considered possible candidates. But they do not license IP-preposing as shown in (63).

(63) *$[_{IP}$ Bill will visit tomorrow $]_1$, I don't know whether/if t_1.

Independently of examples such as (63), it is dubious that they are in CP Spec agreeing with the head C, since they do not license Sluicing either, as shown in (8b). Another relevant example is given in (64).[15]

(64) *Mary thinks Bill will visit tomorrow, but I don't know whether/if.

We tentatively assume here that *whether/if* are in Comp, and do not agree with an operator in CP Spec. Then, (63)-(64) are straightforwardly ruled out by the condition in (57).

We have seen that (57) has the effect of excluding all cases of IP-preposing. As is clear from its formulation, it excludes some cases of NP-preposing as well. Let us consider (65).

(65) *Book$_1$, I read [the/a t_1].

This example is ruled out by (57), exactly as the example of NP-deletion in (6b), repeated below as (66), since the D dominating *the/a* does not agree with its Spec, and hence, is not a licensing head.

(66) *I read about that person, and now, I want to see the.

Then, what remains to be accounted for are examples like (52a-b), repeated below as (67a-b).

(67) a. *Book$_1$, I read [John's t_1].
b. *Destruction of the city$_1$, I witnessed [the barbarians' t_1].

Neither of these examples are ruled out by (57), since as indicated by the presence of the genitive phrase, the D agrees with its Spec and thus licenses the trace. Further, given our conclusion in Section 2, (67b) need not violate the Proper Binding Condition. More specifically, it can have the structure in (68).

(68) [$_{NP}$ PRO$_1$ destruction of the city]$_2$, I witnessed [$_{DP}$ the barbarians'$_1$ t_2].

The same problem arises with (67a), but independently of our PRO analysis if possessors can be base-generated in DP Spec. In the following subsection, we will suggest an account for these examples.

3.2. NP-Preposing and the ECP

Let us first consider the following examples:

(69) a. *How$_1$ does John wonder [$_{CP}$ who$_2$ [$_{IP}$ t_2 fixed the car t_1]]?

b. ??What$_1$ does John wonder [$_{CP}$ who$_2$ [$_{IP}$ t_2 fixed t_1]]?

Rizzi (1990), proposing Relativized Minimality, attributes the ill-formedness of (69a) to the fact that the operator *who* in CP Spec intervenes between *how* and its trace. To make the discussion more precise, let us assume the representation in (70) for this example.

(70) $[_{CP}$ How$_1$ $[_{\overline{C}}$ does $[_{IP}$ t_1^5 $[_{IP}$ John $[_{VP}$ t_1^4 $[_{VP}$ wonder $[_{CP}$ who$_2$ $[_{IP}$ t_1^3 $[_{IP}$ t_2 $[_{VP}$ t_1^2 $[_{VP}$ fixed the car t_1^1]]]]]]]]]]?

Here, *who* in CP Spec intervenes between t_1^4 and t_1^3, and hence, the former fails to antecedent govern the latter. As a result, the chain of *how* fails to satisfy the antecedent government requirement on each link of a chain, and the example is ruled out.

Chomsky and Lasnik (1993) extends this analysis to the weaker violation in (69b), and at the same time, propose to derive Rizzi's Relativized Minimality from the following principle of Economy of Derivation:[16]

(71) Minimize chain links.

This principle states, roughly, that movement must proceed via every possible landing site. As the movement of *how* in (70) from the position of t_1^3 to that of t_1^4 skips the embedded CP Spec, a possible landing site, this movement is illegitimate. They propose that a trace produced by such illegitimate movement, e.g., t_1^3 in (70), is marked with a *. Given this analysis, the derivation of (69b) also produces a starred trace, as illustrated in (72).

(72) $[_{CP}$ What$_1$ $[_{\overline{C}}$ does $[_{IP}$ t_1^5 $[_{IP}$ John $[_{VP}$ t_1^4 $[_{VP}$ wonder $[_{CP}$ who$_2$ $[_{IP}$ $*t_1^3$ $[_{IP}$ t_2 $[_{VP}$ t_1^2 $[_{VP}$ fixed t_1^1]]]]]]]]]]?

In this example too, t_2^3 is produced by a movement that skips a possible landing site and fails to satisfy (71).

In order to account for the contrast between (69a) and (69b), Chomsky and Lasnik appeal to the analysis of the argument/adjunct asymmetry outlined in Chomsky (1991). Suppose that deletion, like movement, applies only as the "last resort," and further, that the chains in (73a-c) but not the one in (73d) constitute legitimate objects at LF.

(73) a. A ... A ... A (uniform A-chain)
b. A' ... A' ... A' (uniform A'-chain)
c. A .... A (operator-variable chain)
d. *A' ... A' ... A

Then, the trace t_1^3 in (70) cannot delete in LF, since it is a part of a legitimate chain of the form (73b). On the other hand, the corresponding trace in (72) must, and therefore can, delete in LF along with the other intermediate traces. Such deletion is necessary to create an operator-variable chain from an illegitimate chain of the form (73d). Given this distinction, Chomsky and Lasnik propose that the creation of a starred trace constitutes a minor violation, while the presence of such an "offending" trace at LF results in much stronger deviance (their version of the ECP).

The analysis outlined above directly extends to the examples of subject extraction in (74).

(74) a. Who$_1$ does John think [$_{CP}$ t_1' [$_{IP}$ t_1 fixed the car]]?

b. *Who$_1$ does John wonder [$_{CP}$ how$_2$ [$_{IP}$ t_1 fixed the car t_2]]?

Let us suppose, following Frampton (1990), that the IP-adjoined position is not a possible landing site for the subject immediately dominated by the IP. We state the relevant constraint as follows:

(75) a. A chain link must be at least of length 1.
b. A chain link from A to B is of length n iff there are n "nodes" (X, $\overline{X}$ or XP, but not segments of these) that dominate A and exclude B.

(75a) states basically that each chain link must have some length. It excludes the chain link from the IP Spec to the IP-adjoined position, since according to (75b) the chain link is of length 0. A constraint of this kind seems to be needed independently: without it, (71) may force a phrase in an adjoined position to keep adjoining to the same node.[17]

Given (75), *who* in (74b) must skip the CP Spec when it creates the first link of the chain. The movement of *who* is illustrated in (76).

(76) [$_{CP}$ Who$_1$ [$_{\overline{C}}$ does [$_{IP}$ t_1^3 [$_{IP}$ John [$_{VP}$ t_1^2 [$_{VP}$ wonder [$_{CP}$ how$_2$ [$_{IP}$ *t_1^1 fixed the car t_2]]]]]]]]?

Here, unlike the case of object extraction, the initial trace is assigned a *. Since this trace cannot delete, we correctly predict a severe violation. Note that *who* in (74a) can move through the embedded CP Spec, and hence, no starred trace needs to be created.

This analysis works well also for extraction out of DPs. Let us consider the following examples from Stowell (1989):

(77) a. Who$_1$ did you sell [$_{DP}$ [$_{\overline{D}}$ a [$_{NP}$ picture of t_1]]]?

b. Who$_1$ does Jane regret [$_{DP}$ [$_{\overline{D}}$ the [$_{NP}$ dismissal of t_1]]]?

(78) a. ??Who$_1$ did you sell [$_{DP}$ Mary's [$_{\overline{D}}$ [$_{NP}$ picture of t_1]]]?

b ??Who$_1$ does Jane regret [$_{DP}$ Bob's [$_{\overline{D}}$ [$_{NP}$ dismissal of t_1]]]?

As discussed in detail in the literature, a genitive phrase blocks extraction out of a DP. (See, among others, Torrego (1986), Stowell (1989), Ormazabal (1991), and the references cited there.) According to the standard analysis originally proposed by Torrego, the examples in (77) are well-formed since *who* can move through the DP Spec position, while those in (78) are out because such movement is blocked by the genitive phrase. And as should be clear, this analysis can be stated quite naturally in terms of "Minimize chain links." Let us consider the following derivation for (78a):

(79) $[_{CP}$ Who$_1$ $[_{\bar{C}}$ did $[_{IP}$ t_1^4 $[_{IP}$ you $[_{VP}$ t_1^3 $[_{VP}$ sell $[_{DP}$ Mary's $[_{\bar{D}}$ $[_{NP}$ $*t_1^2$ $[_{NP}$ picture of t_1^1]]]]]]]]]]?

As *who* skips the DP Spec, a possible landing site, the trace t_1^2 is marked with a *. But being an intermediate trace of an argument wh-phrase, this trace must, and therefore can, delete in LF. Thus, we have a weak violation. Since the movement in (77) can go through the DP Spec, no possible landing site needs to be skipped, and hence, no trace needs to be starred.[18]

Let us now return to the problematic examples of NP-preposing in (67), repeated below in (80).

(80) a. *Book$_1$, I read [John's t_1].
b. *Destruction of the city$_1$, I witnessed [the barbarians' t_1].

Note first that the analysis of (78) basically carries over to these examples. As it was assumed above that DP Spec is a possible landing site for wh-movement, it is reasonable to suppose that it is a possible landing site for NP-preposing as well. And the movement in (80a-b) clearly skips this possible landing site.[19] The derivation of (80a) is illustrated in (81).

(81) $[_{IP}$ $[_{NP}$ book$]_1$ $[_{IP}$ I $[_{VP}$ t_1' $[_{VP}$ read $[_{DP}$ John's $[_{\bar{D}}$ $[_D$ $\underline{e}$]] $*t_1$]]]]].

Here, the trace t_1, like t_1^2 in (79), is marked with a * since it was created by an illegitimate movement skipping the DP Spec position. At the same time, however, there is one important difference between (78) and (80). In the latter, but not in the former, the initial trace is starred, as is clear from the structures in (79) and (81). Hence, in the latter case, the offending trace remains at LF, and a stronger violation is expected. Thus, the theory of Chomsky and Lasnik (1993) correctly accounts for the three-way contrast among (77), (78), and (80).

We have seen in this subsection that Chomsky and Lasnik's (1993) ECP, which was originally proposed, in part, to account for the argument/adjunct asymmetry in (69), has the desirable effect of ruling out potentially problematic examples of NP-preposing. Note that it also excludes (redundantly) the examples of IP-preposing such as (62), repeated below in (82).

(82) *$[_{IP}$ John bought t_2 $]_1$ I don't know $[_{CP}$ what$_2$ $[_{\bar{C}}$ $[_C$ $\underline{e}$] t_1]].

The IP-preposing here skips the CP Spec and hence, must leave a starred initial trace, exactly as in (81). More generally, Chomsky and Lasnik's ECP prohibits

NP/IP-preposing whenever the DP/CP Spec is filled. Then, all cases of NP/IP-preposing are excluded since unless the DP/IP Spec is filled, the trace of NP/IP fails to be licensed, according to the Head Licensing Condition discussed in the preceding subsection.

The situation is different in the case of VP-preposing. In this case also, the Head Licensing Condition requires that the IP Spec be filled. But this does not result in a starred initial trace of the proposed VP, because the IP Spec, being strictly an A-position, is not a possible landing site for the VP. What makes VP-preposing possible, ultimately, is then the fact that IP Spec is strictly an A-position.

4. CONCLUSION

In Section 2 of this paper, we argued for the possibility of the structures in (83).

(83) a. $[_{IP}$ The barbarians$_1$ $[_{\bar{I}}$ I $[_{VP}$ PRO$_1$ $[_{\bar{V}}$ destroyed the city]]]].

b. $[_{DP}$ The barbarians'$_1$ $[_{\bar{D}}$ D $[_{NP}$ PRO$_1$ $[_{\bar{N}}$ destruction of the city]]]].

This of course does not preclude raising within IP and DP. Our conclusion is that IPs and DPs are ambiguous with respect to raising and control. It is well known that modals such as *can*, *may*, and *must* have two interpretations, "root" and "epistemic," as illustrated in (84).

(84) John can win the race
a. John has the ability to win the race (root)
b. It is possible that John will win the race (epistemic)

And it seems reasonable to assume that the "root" *can* takes a base-generated subject, while the subject of the "epistemic" *can* is raised from a VP-internal position. As Jim Huang points out (p.c., 1986), our proposal, then, amounts to saying that IP and DP are in general structurally ambiguous in this way.

In Section 3, we examined why IP-preposing and NP-preposing, as opposed to VP-preposing, are in general impossible. We first showed that all cases of IP-preposing are excluded by the Head Licensing Condition. Then, noting that this account does not extend to some cases of NP-preposing, we proposed an analysis for those cases in terms of Chomsky and Lasnik's (1993) ECP. According to this analysis, the cases of NP-preposing that satisfy the Head Licensing Condition are excluded in the same way as the adjunct/subject extraction over a wh-island in (69a) and (74b): "Minimize chain links" forces a starred trace and the offending trace remains at LF. This analysis, if correct, provides support for the ECP, as formulated in Chomsky and Lasnik (1993). The discussion in this section shows that the asymmetry between VP and NP/IP with respect to preposing is straightforwardly explained with the assumption that IP, DP, and CP have completely parallel structures. In this sense, it supports the parallel analysis of these functional projections.

NOTES

* This is a slightly revised version of the original written in 1993. The material in this paper was presented at the 1989 LSA Annual Meeting (Section 2) and also in syntax seminar at the University of Connecticut. We would like to thank Sung-Ho Ahn, Mona Anderson, Joseph Aoun, Zeliko Bošković, Noam Chomsky, Naoki Fukui, Jim Huang, Ray Jackendoff, Kyle Johnson, Howard Lasnik, Rhang Lee, Roger Martin, Javier Ormazabal, and Daiko Takahashi for helpful comments and suggestions. The first author started working on the material in Section 2.3 when he was at the University of Southern California in 1984-86. There, he benefitted greatly from discussions with Osvaldo Jaeggli on this material.

[1] The works mentioned above offer different hypotheses with respect to the exact nature of the VP-internal subject position, and also with respect to the positioning of the genitive *'s*. But we will not be concerned with those differences, since as far as we can see, the discussion in this paper is consistent with any of the alternative hypotheses.

[2] In an early version of Lebeaux (1983), he entertains the hypothesis that the subject of a finite clause controls PRO in Infl, and thus, is licensed via predication. See also Hyams (1985) for much relevant discussion.

[3] See Martin (1992) for a detailed discussion. As he notes, examples such as the following constitute evidence for the null Case hypothesis for PRO:

(i) John wants me to take the exam, but I don't want [PRO to].

If PRO agrees with Infl and receives null Case, as hypothesized in Chomsky and Lasnik (1993), then the well-formedness of (i) is exactly what we expect, given the generalization in (5b). (See also Saito and Murasugi (1990).)

We tentatively assume here that PRO can be assigned null Case, but does not have to have Case to receive a Θ-role, for reasons that will become clear shortly.

[4] Similarly, it seems that adjuncts are unable to move to IP Spec, as illustrated below.

(i) a. It seemed then that John was going to win the prize.
b. *Then$_1$ seemed t_1 that John was going to win the prize.

[5] Here, the PRO in (20) must be in NP Spec, and cannot be in DP Spec. If it raises from NP Spec to DP Spec, then it c-commands John. Thus this raising creates a configuration in violation of Condition (C).

[6] Jackendoff was not advocating the structure in (26) itself, but he showed us many examples like (28)-(29) which cast doubts on the NP-deletion analysis of Anderson's paradigm, proposed in Saito and Murasugi (1990).

[7] We unfortunately noticed this fact after the final draft for Saito and Murasugi (1990) was completed. Ray Jackendoff (p.c., 1989) has also pointed out this fact in his comments on Saito and Murasugi (1990).

[8] The argument below for (36) is presented also in Ahn (1990) only with minor differences.

[9] The relevance of examples such as (41a-b) to the Proper Binding analysis of (39) was first pointed out to us by Joseph Aoun and Kyle Johnson (p.c., 1986).

[10] The potentially problematic examples alluded to above are the following:

(i) a. They said the cat would be let out of the bag, and let out of the bag, it (the cat) was.
b. ?They said advantage would be taken of him, and taken of him, it (?*advantage) was.

These examples are particularly mysterious, since the parallel raising examples are out, as shown in (ii) and (46).

(ii) *[How likely to be let out of the bag $]_1$ is the cat t_2?

It is not clear to us at this point why (ia) and (ii), for example, have a completely different grammatical status.

[11] Note that this difference between NP and VP requires an analysis independently of the structures in (50) we argued for. Even under the standard VP/NP-internal subject hypothesis with a trace in the VP/NP-internal subject position, it is not obvious why (52b) should be out when VP-preposing is possible. And if possessor genitives can be base-generated in DP Spec, as we have been assuming, then the ungrammaticality of (52a) should be explained independently of the Proper Binding Condition.

[12] See also Jaeggli (1985), Stowell (1986), Aoun, Hornstein, Lightfoot and Weinberg (1987), Rizzi (1990), and Chomsky and Lasnik (1993) for discussions on this condition.

[13] We revised the name of the condition, since it is not clear to us that the notion "government" plays any role in the formulation of this condition. See Bošković (1993) and Saito (1992) for relevant discussion. As it is not clear that this condition should apply to "intermediate traces," it may be necessary to construe it as a condition on empty categories in the tail position of a chain.

[14] This seems incompatible with Rizzi's (1990) analysis of the *that*-trace effect, adopted in Chomsky and Lasnik (1993), and also with the analysis suggested in Lasnik and Saito (1992). For an alternative analysis of the *that*-trace effect, see, among others, Takahashi (1993).

[15] It seems clear that the Japanese counterpart of *whether/if* occupies the Comp position. The language is head-final/Spec-initial, and *kadooka*, which corresponds to *whether/if*, follows the complement IP. As shown in Takahashi (1994), *kadooka* does not license Sluicing, as expected.

[16] See Takahashi (1993) for a detailed discussion, as well as an extension, of this principle.

[17] See Bošković (1994) for much relevant discussion. It is reasonable to suppose that (75a) should be derived from a general principle of economy. Chomsky's "no superfluous steps" seems to be a natural candidate.

[18] As discussed in Huang (1982) and the references cited in the text, extraction of an adjunct out of a DP uniformly results in a strong violation. The following example from Stowell (1989) illustrates this fact:

(i) *By whom$_1$ did Frans sell [a picture t_1]?

Once a starred trace is created in the derivation, we expect a strong violation, since the chain is a uniform A'-chain and hence no trace can be deleted. The desired result is obtained if we assume, as discussed in Section 2, that adjuncts, as opposed to arguments, cannot move to DP Spec. This will force the movement to skip a possible landing site, i.e. DP Spec, and create a starred trace.

[19] The DP Spec position is a possible landing site for the NP, according to (75), since the resulting chain link will be of length 1: the $\overline{\text{D}}$ node dominates the NP and excludes the DP Spec position.

REFERENCES

Abney, S.: 1986, 'Functional Elements and Licensing', Presented at the 1986 GLOW Conference, Barcelona.

Ahn, S.-H.: 1990, *Korean Quantification and Universal Grammar*, Doctoral Dissertation, University of Connecticut.

Anderson, M.: 1983, 'Prenominal Genitive NPs', *The Linguistic Review* 3, pp.1-24.

Aoun, J., N. Hornstein, D. Lightfoot and A. Weinberg: 1987, 'Two Types of Locality', *Linguistic Inquiry* 18, 537-577.

Barss, A.: 1986, *Chains and Ananhoric Dependence*, Doctoral Dissertation, MIT.

Bošković, Z.: 1993, 'Clausal Selection, Minimality, and Subjacency', ms., University of Connecticut.

Bošković, Z.: 1994, 'D-Structure, Θ-Criterion, and Movement into Θ-Positions', *Linguistic Analysis* 24, 247-286.

Chomsky, N.: 1991, 'Some Notes on Economy of Derivation and Representation', In R. Freidin, ed., *Principles and Parameters in Comparative Grammar*, MIT Press, Cambridge, Mass., 417-454.

Chomsky, N. and H. Lasnik: 1993, 'The Theory of Principles and Parameters', In J. Jacobs, et al., eds., *Syntax: An International Handbook of Contemporary Research*, Walter de Gruyter, Berlin, pp.506-569.

Frampton, J.: 1989, 'Parasitic Gaps and the Theory of Wh-Chains', *Linguistic Inquiry* 21, 49-77.

Fukui, N. and M. Speas: 1986, 'Specifiers and Projection', *MIT Working Papers in Linguistics* 8, 28-172.

Giorgi, A. and G. Longobardi: 1991, *The Syntax of Noun Phrases*, Cambridge University Press, Cambridge.

Huang, C.-T. J.: 1982, *Logical Relations in Chinese and the Theory of Grammar*, Doctoral Dissertation, MIT.

Hyama, N.: 1985, *Language Acquisition and the Theory of Parameters*, Doctoral Dissertation, CUNY Graduate Center.

Jackendoff, R.: 1971, '"Gapping and Related Rules', *Linguistic Inquiry* 2, 21-35.

Jaeggli, O.: 1982, *Topics in Romance Syntax*, Foris Publications, Dordrecht.

Jaeggli, O.: 1985, 'On Certain ECP Effects in Spanish', ms. University of Southern California.

Kitagawa, Y.: 1986, *Subjects in Japanese and English*, Doctoral Dissertation, University of Massachusetts, Amherst.

Koopman, H. and D. Sportiche: 1985, 'Theta-Theory and Extraction', Presented at the 1985 GLOW Conference, Brussels.

Kuroda, S.-Y.: 1988, 'Whether We Agree or Not: A Comparative Syntax of English and Japanese', In W. Poser, ed., *Japanese Syntax*, CSLI Publications, Stanford.

Lasnik, H. and M. Saito: 1992, *Move-α: Conditions on its Application and Output*, MIT Press, Cambridge, Mass.

Lebeaux, D.: 1983, 'Locality and Anaphoric Binding', ms. University of Massachusetts, Amherst.

Lobeck, A.: 1990, 'Functional Heads as Proper Governors', NELS 20, 348-362.

Martin, R.: 1992, 'On the Distribution and Case Features of PRO', ms. University of Connecticut.

Ormazabal, J.: 1991, 'Asymmetries in WH-Movement and Some Theoretical Consequences', ms. University of Connecticut.

Rizzi, L.: 1990, *Relativized Minimality*, MIT Press, Cambridge, Mass.

Saito, M.: 1986, 'LF Effects of Scrambling', Presented at the First Princeton Workshop on Comparative Grammar.

Saito, M.: 1989, 'Scrambling as Semantically Vacuous A'-Movement', In M. R. Baltin and A. S. Kroch, eds., *Alternative Conceptions of Phrase Structure*, University of Chicago Press, Chicago, 182-200.

Saito, M.: 1992, 'On the Head Government Requirement on Empty Categories', Presented at the Spring 1992 Meeting of the English Literary Society of Japan.

Saito, M. and K. Murasugi: 1990, '$\bar{N}$-Deletion in Japanese', *UConn Working Papers in Linguistics* 3, 87-107.

Stowell, T.: 1986, 'Null Antecedents and Proper Government', *NELS* 16, 476-493.

Stowell, T.: 1989, 'Subjects, Specifiers, and X-Bar Theory', In M. R. Baltin and A. S. Kroch, eds., *Alternative Conceptions of Phrase Structure*, University of Chicago Press, Chicago, 232-262.

Takahashi, D.: 1993, 'Minimize Chain Links', Presented at the Spring 1993 Meeting of the English Literary Society of Japan.

Takahashi, D.: 1994, 'Sluicing in Japanese', *Journal of East Asian Linguistics* 3, 265-300.

Torrego, E.: 1986, 'On Empty Categories in Nominals', ms. University of Massachusetts, Boston.

Zagona, K.: 1982, *Government and Proper Government of Verbal Projections*, Doctoral Dissertation, University of Washington.

SUBJECT CLITICS IN FRENCH AND ROMANCE COMPLEX INVERSION AND CLITIC DOUBLING*

Dominique Sportiche
University of California, Los Angeles

The analysis of the French construction of Complex Inversion illustrated in (1) below raises difficulties with respect to current conceptions of available phrase structure:

(1) a. Jean est-il malade?
'Is John sick?'

b. Depuis quand Jean est-il malade.
'Since when is John sick.'

The problem arises from the conjunction of the following several factors. First, complex inversion is a root phenomenon. It is consequently reasonable, as suggested generally by Den Besten extrapolating from the classic V2 effect and adopted ever since, to suppose that it involves raising of I to C. Secondly, Wh-movement is to [Spec,CP]. Third, it appears that the subject DP intervenes between the wh-phrase and the highest verb (as in (1)). Where then is this DP subject? Finally, and to compound the problem, it appears that a pronominal copy of the subject may cooccur with a full DP subject, making complex inversion a construction with two subjects.

In this paper, I will propose an analysis of this construction (as well as of the simpler subject clitic inversion - *est-il malade/is he sick*) that builds on and complements two theoretical developments.

First, the theory of clitic constructions developed in Sportiche (1995a,c, to appear) which focused primarily on object clitics. I will show that this theory naturally extends to Subject Clitics (SCL). I will also show that it leads to simple analyses of Subject Clitic Inversion (SCI) and of Complex Inversion (CI), making Complex Inversion much less exceptional than other proposals have made it, in that no structure building is required. I will further suggest it entails a much greater deal of underlying similarity across Romance languages than most analyses have assumed.

Second, the conception of morphology alluded to in Chomsky (1992): this conception is meant to provide a principled account of the cases of affix lowering, whereby morphological concatenation of an inflectional morpheme with some other unit is not

K. Johnson and I. Roberts (eds.), Beyond Principles and Parameters, 189-221.

a syntactic operation, but checking of the properties of this inflectional morpheme is done through the syntactic operation of head movement.

1. SOME BACKGROUND

The approach to clitic constructions suggested in Sportiche (1995a) can be summarized as follows: In a Romance clitic construction, there is an inflectional head H^O heading a projection within the inflectional system, call it ClP in the general case. I will take this H^O to be the clitic itself generally.[1]

Thus, a sentence like (2) receives the analysis in (3):

(2) Marie les_1 aura présentés XP_1* à Nick.
them would-have introduced-AGR_1 to Nick
Marie would have introduced them to Nick

(3) $[_{AccP}$ DP^ $[[_{Acc}$les] [... aura présentés DP^*_{+acc} ...]]]

The XP associated with this clitic, that we will refer to as XP* throughout (here a DP* analyzed as *pro*) must raise to the Specifier position (here DP^) of ClP (here accusatively marked or AccP) by LF, thereby guaranteeing agreement between the two and explaining why the clitic/XP* relation (in fact the XP^/XP* relation) can only occur between two entities over a syntactic context transparent to movement. In French, Raising of XP* to XP^ takes place overtly (as shown by the triggering of object agreement - see Sportiche (1995a) for detailed discussion).

This XP* may be required to be covert (*pro*) as in French, or may be overt, yielding various instances of clitic doubling. If it is overt, movement to XP^ may be overt too yielding various cases of Scrambling (in the Germanic languages) or involved in Clitic Left Dislocation constructions.

If there are several clitics, there are as many ClPs. The function of these ClPs is not necessarily uniform. Although clitics share a number of constructional properties, they may diverge on others as e.g. the interpretive consequences of their presence. Thus typical accusative clitics induce specificity of their associated XP while dative clitics do not. In Sportiche (1995a), I have suggested that the function of Accusative ClP in Romance is to license Specificity and in this is a projection different from AGR_oP the function of which is to license Case assignment, when available. The function of the Dative ClP does seem akin to AGR_oP in licensing Dative Case and would thus best be analyzed as $AGR_{io}P$. Correlated with these differences is the A-bar status of the Specifier position of the accusative ClP, making it analogous to an operator projection in keeping with its interpretive function. The Specifier of $AGR_{io}P$ on the other hand is an A-position, as we would expect by analogy with the Specifier of the agreement projections licensing Accusative or Nominative.

2. QUESTIONS ABOUT SUBJECT PRONOUN CLITICS

The set of elements referred to as pronominal clitics found in French is illustrated by the following template:

(4) French Clitic Template

Nom	1st/2nd/Refl	3rd Acc predicate	3rd Dat	Loc	Gen
il	me/te/se/nous	le/la/les	lui/leur	y	en

As Perlmutter (1972) describes it, for each column of the template, no more than one clitic is allowed (so for example, it is not possible to have a 1st person accusative object and a 2nd person dative object at the same time). Furthermore, the third and fifth column cannot be filled at the same time (e.g. * *Je me lui montre / I show myself to him*). These restrictions are discussed to a limited extent in Sportiche (1995a). For the most part, their existence and properties are not understood.

Clitics are so called because they are claimed not to constitute autonomous syntactic units. Rather, they seem to form a unit with some host either syntactically (syntactic clitics) or phonologically (phonological clitics).[2] Henceforth, clitic without further specification always means syntactic clitic. Kayne (1975) argues that object pronominal clitics adjoin to V in French (a conclusion with substantial cross Romance validity). This is exemplified by the fact that they cannot be conjoined independently of their V host nor can their V host be conjoined independently of them. Disjunction is however more acceptable at least for 3rd pers items of the same column in the template.

(5) a. *Jean le et les présentera a Pierre.
'John will introduce him and them to Peter.'

b. *Jean le présenté a Pierre et/ou envoie a Paris.
'John presents him to Peter and/or sends him to Paris.'

c. ?Pierre le ou les (le ou la) verra au concert.
'Pierre him or them (him or her) will see at the concert.'

Other properties typical of object clitics are their lack of stress and the impossibility to modify them by an adjacent modifier.

The subject "pronouns" *je, tu, il, ils, on (I, you-sg, he, they-mpl, one*) only occur as nominative. The other pronominal subjects *elle, elles, nous, vous /she, they-fpl, we, you-pl* are not exclusively nominative. They may occur for example as the object of prepositions (e.g. *avec *il/*tu/elle/nous (with he/you/she/we*). Like objects clitics, subject clitics disallow being conjoined, i.e. the equivalent of (5a) is unacceptable as shown in (6a). Surprisingly, definite subject proforms allow the equivalent of (5b), i.e. their verbal host may be conjoined independently of them. The indefinite subject proform *on* (which we imperfectly gloss as "one" here - see Cinque, 1988, for relevant discussion), as well as the proform *ce*,[3] disallow this option. These facts are illustrated in (6b). Disjunction as in (6c) seems more acceptable for subject forms than for object forms:

(6) a. *Il/tu/je.. et tu/je/elle... connaiss-ont/ez/ent Marie.
'He and she know Marie.'

b. Il/tu/*on mangé du pain et boit du vin.
'He/you/one eats bread and drinks wine.'

c. Il ou elle connait Marie.
'He or she knows Marie.'

Weak object pronouns clearly are clitics. If it is correct that they head an inflectional projection as described above, they are not pronominal in the sense that they do not stand for a regular DP. The same questions arise for subject clitic pronouns:

What is their categorial status?
Where in the structure are they generated?
Are they syntactic clitics and/or phonological clitics?

Furthermore, since subject clitics may either appear preverbally (e.g. in statements) or postverbally (in SCI or CI constructions), the last question arises for both cases. To anticipate the conclusions I will reach, I will argue that preverbal subject forms conform to the general picture of clitics argued for in Sportiche (1995a,c, to appear) and described above: they are generated as heads of an Inflectional-like category, call it NomP, and not of a DP; they are located above AGRsP in the structure of a clause; they are not necessarily, although they may be, clitics (except for *on, ce* which always are clitics). Postverbal clitics on the other hand always are clitics and not DPs. I will suggest that they be treated as morphological affixes base generated on the verb itself and checked in the head position of NomP (much the way tense affixes are generated on the verb in English and checked in T).

3. EXTENDING THE ANALYSIS TO SUBJECT CLITICS (SCL)

Suppose we extend the analysis offered in Sportiche (1995a) for Accusative clitics to Nominative clitics. What then would we expect? Consider a sentence with an object clitic:

(7) Jean l'a repeinte.
'John has repainted it(FEM).'

We analyze its relevant aspects as follows (AGRs omitted, see Sportiche (1995a) for details):

(8) Jean T _ Cl avoir _ AGR_o peint *pro*
ACC

The function of AGR_o is to license Accusative Case. Because in French, *pro* raises to [Spec,ClP] in the syntax, (optionally) transiting through [Spec,AGR_o], object agreement is (optionally) triggered on the participle (see Sportiche (1989)). The head of ClP, *le,* ends up cliticized to *avoir* which has moved to T.

For a nominative clitic, we would expect two different projections, one responsible for subject agreement and sanctioning Nominative Case, and one for the clitic. Furthermore, given that the clitic is the head of the Nominative ClP, it should a priori be able to cooccur with an overt DP* in its Specifier or with a covert one.

4. TRENTINO ITALIAN

The former expectation is actually fulfilled as (a slight modification[4] of) the analysis of subject clitics in some Northern Italian dialects presented in Rizzi (1986) or Brandi and Cordin (1989) illustrates. Consider for example Trentino as a representative of such dialects (which include at the very least Fiorentino):

(9) El Gianni/O el magna.
'Gianni/he eats.'

As Rizzi (1986) or Brandi and Cordin (1989) discuss in a pre-split-Infl framework, the subject clitic occupies the head of an inflectional projection (namely AGR in Infl for them) and the DP subject the regular subject position as follows:

(10) [El Gianni $[_{\text{Infl}}$ el] $[_{\text{VP}}$ magna]].
Gianni (he) eat-PRES-3SG

This analysis is inadmissible within the more restrictive current theoretical framework: this Infl projection cannot stand for the usual subject agreement. The property of (unsplit) Infl includes both Tense specification and Subject/verb agreement. These properties do show up morphologically on the verb independently of the subject clitic exactly as they do in standard Italian or French. This suggests that this inflectional projection headed by *el* is an additional inflectional projection. We can immediately make sense of its existence within our general proposal for clitics by identifying this Infl node with Nominative ClP. Taking into account subject verb agreement (and ignoring Tense, which would, if represented within the split Infl paradigm, require an additional TP projection lower than AGRs), we are led to enrich the structure as follows:

(11) $[_{\text{NomP}}$ El Gianni_1/pro_1 $[_{\overline{\text{Nom}}}$ $[_{\text{Nom}}$ el] $[_{\text{AGRs}}$ t_1 ...magna ...]]].

The DP subject has raised through [Spec,AGRs] getting Nominative Case, and in [Spec, NomP] (where NomP is ClP with Cl in the nominative), the clitic heads NomP and the verb is in or is connected to AGRs thereby agreeing with the subject. This result is the exact analogy for nominative clitics of our conclusions for accusative clitics. In our view, then, the syntactic structure of Trentino subject clitics fits the general mold of the analysis of "pronominal" clitics (as in, and in the form argued in, Sportiche (1983) or in Roberge (1990)).[5]

5. FRENCH PREVERBAL SUBJECT CLITICS

Can French preverbal subject clitics be analyzed along similar lines as Trentino's. If the analysis of Romance clitics is fully general, we expect a positive answer. Prior to Rizzi's or Brandi and Cordin's work on the northern Italian dialects, this was actually a dominant approach to the analysis of preverbal subject clitics. For example, Jaeggli (1982), Sportiche (1983) (or more recently Roberge, 1990) argue that they are generated as part of an complex Infl node.

However, both Rizzi and Brandi and Cordin argue against this suggestion. Rather, they argue that French subject clitics head DPs, i.e. are generated in the same position as full DP subjects. I will now argue that a modernized version of Jaeggli's analysis along the lines of the general analysis of clitics I offer is superior, i.e. that French can and should be analyzed like Trentino is above. In this, our analysis is extremely close to Safir's (1982) or Roberge's (1990).

Rizzi (1986), and Brandi and Cordin (1989), henceforth RBC, propose the following analysis for French and Trentino respectively (where Cl is the underlying position of the SCL):

(12) Trentino French

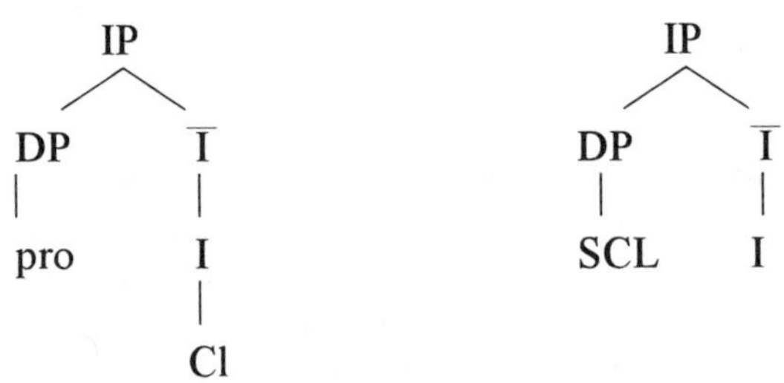

The logic of their reasoning is as follows: there are systematic differences (mostly distributional, e.g. with respect to Complex Inversion) between northern Italian subjects clitics and French subject clitics: since Northern Italian clitics sit in Infl (i.e. is subject agreement for them), French subjects clitics cannot, and must thus be heads of the subject DP (which may eventually cliticize, as Kayne (1972; 1975 or 1984) has suggested). I will show that their correct observations do not entail their conclusions and are completely compatible with an analysis of French along the lines of that of Trentino.

They consider three differences between Trentino and French:

(i) Standard French SCLs, unlike Trentino's, cannot cooccur with a full DP subject.
(ii) Some but not all French SCL may cooccur with a conjoined VP. Trentino SCL may not.
(iii) French subject clitic paradigms are regular, unlike those of the northern Italian dialects.

Let us discuss them in turn.

5.1. Complementarity of Distribution

Let us begin with the first difference: Standard French SCL cannot cooccur with a full DP subject. Some dialects of French allow cooccurrence but never with the full range of DPs that Trentino SCL allows. Furthermore, Trentino SCL are obligatorily present if possible, French SCL are not:

(13) a. */OK Jean il mange / Jean mange /il mange.
'John he eats /John eats / he eats.'

b. El Gianni el magna/ *El Gianni magna/ el magna.
'John he eats/ John eats/ he eats.'

c. *Personne il n'a rien dit.
'Nobody he said anything.'

d. Nessuno l'ha detto nulla.
'Nobody he said anything.'

Note first, as indicated, that (13a) is acceptable - without dislocation intonation - for certain French speakers (e.g. from Quebec).[6] As noted by Rizzi, Brandi and Cordin, this distributional complementarity found in standard French is accounted for if the French SCL is the head of the DP subject: the clitic and the full subject would then be competing for the same position. Although this kind of explanation is plausible for cases in which two identical elements - as e.g. two heads, or two phrasal categories - are in complementary distribution, it is considerably weakened by the fact that complementarity is between a clear phrasal category - the subject - and a single word - the clitic. Indeed, it is reminiscent of other cases in which competition for the same position is less plausible: consider for example the doubly filled CP filters.[7] This is a case of complementarity between a word (the head C of CP) and a phrase (a wh-phrase in [Spec,CP]). There is no a priori reason to believe that the case of complementarity under discussion is of the first type (competition for the same position) rather than of this second type (a doubly filled phrase prohibition). This second type is of course immediately compatible with our proposal: if the French SCL heads a nominative ClP, i.e. NomP with the subject DP in [Spec,NomP], the configuration would be ruled out by a doubly filled NomP filter prohibiting the head and the Specifier of the nominative ClP from being lexically filled simultaneously in standard French. This filter would not operate in Trentino nor in certain dialects of French. Additionally, note that the complementarity found for SCL and its account exactly parallels the data for object Clitics and the account offered in Sportiche (1995a): French Object Clitics (OCL) also are in complementary distribution with their associated XP*. It turns out then that this doubly filled ClP prohibition is fully general in standard French. Furthermore, handling those French dialects in which this prohibition does not hold for SCL is straightforward, and does not require analyzing SCL any differently from the way they are analyzed in standard French.

Of course, the preceding discussion only shows that the account we just outlined is a reasonable alternative to the one proposed in Rizzi (1986). We now turn to some independent evidence in its favor.

As in the case of wh-movement to [Spec,CP], the doubly filled ClP (resp. CP) prohibition does not hold if the Specifier of the ClP (resp. CP) contains a trace or a *pro* instead of a lexical element.

(14) Who do you think *t* that John saw?

This situation with ClP is illustrated by Q-float structures. Consider the following sentences (read with continuous non-dislocated intonation):

(15) a. Il a tous fallu qu'il les achètent .
'It is all necessary that he buys them.'

b. Il a tous fallu qu'ils partent.
'It is all necessary that they leave.'

The stranded Q *tous* appears in the main clauses but relates to the object or the subject of the embedded clause. A straightforward account of these data can be constructed if the SCL and OCL are analyzed as heading ClP and not the DP subject or object:

(16) $[\text{tous pro}]_2$.... [ClP t_2 $[_{Cl}$ ils/les]...]

The presence in the main clause of the leftward moved quantifier, an operation only possible via A-bar movement of a phrase, suggests that this Q is part of an XP moved out of its clause.

This XP is most simply and most plausibly analyzed as the subject (or the object) DP itself. Its structure may be *[tous pro]*[8], which is a possible DP as shown by *Tous sont venus* (see Sportiche (1983; 1988) for discussion). Of course, this implies that the SCL cannot be part of this subject DP.

Alternatively this XP may be *[tous t]*, the trace of the SCL cliticized prior to the leftward extra-sentential movement of the Q. In this case, the "underlying" structure *[tous ils]* would have to be postulated, *ils* heading the DP modified by *tous*. This would leave unexplained the ungrammaticality of:

(17) * Tous ils portent / Ils tous sont partis .
'All they left / They all have gone.'

Furthermore, in this case the proposed *[tous t]* in (15b) would contain an unbound trace. This potentially offending trace would require reconstruction of this DP into the lower clause at LF. It is clear however that these leftward moved Q's have matrix scope, not embedded scope. This illustrated by the following pair:

(18) a. Il aurait tous fallu que tu ne les aies pas vu.
'It would have been necessary that you see none of them.

b. Il aurait fallu que tu ne les aies pas tous vu.
'It would have been necessary that you do not see all of them.'

As the paraphrases indicate, in the first sentence the Q must have scope wider than the embedded negation. If reconstruction was required, we would expect the Q to be able to take narrower scope than the embedded negation as in the second sentence. We conclude that the proposed DP cannot be of the form *[tous t]*, hence that the SCL (or the OCL) cannot be part of the subject (resp. object) DP.

Another potential source of problems for the SCL as subject assumption is the existence of clitic doubling structures in questions: complex inversion does allow what appears to be a full subject and a SCL: *Jean est-il malade*, which we will discuss later in section 7.

5.2. SCLs and Conjunction

Let us turn now to the second difference: some but not all French SCLs may cooccur with a conjoined VP. Trentino SCLs may not.

(19) a. Il/*on chante et dense .
'He/one sings and dances.'

b. Il/*on mange de la viande et boit du vin.
'He/one eats meat and drinks wine.'

c. la canto e *(la) balls.
'She sings and (she) dances.'

There is obviously a difference between *il* on the one hand and Trentino *la* on the other. Is this difference an indication of different categorial status? Rizzi (1986) concludes so.

Let us summarize briefly his treatment: he suggests that all these cases of conjunction are ambiguous between clausal conjunction with a null SCL in the second conjunct and VP conjunction.

(20) a. [SCL VP] and [[$_{SCL}$ e] VP]
b. SCL VP and VP

In Trentino, repetition of the SCL is forced because (i) either it is VP conjunction; then, the clitic character of the SCL forces it to cliticize in PF to the first member of the conjunction, violating the Coordinate Structure Constraint, or (ii) it is clausal conjunction but a general principle prohibits a zero *pro* form (here the null SCL) to survive if it is a minor category, but not if it is a major category.

In French, conjunction of VPs is disallowed, bccause SCL are always PF clitics and cliticization would violate the Coordinate Structure Constraint. Clausal conjunction on the other hand is possible because the null SCL in the second conjunct is a major category, a DP in modem terms. Left to explain is the behavior of French *on*, akin to Trentino *la* and unlike French *il*. Here, Rizzi suggests that it is due to the inability of French *on* to act as a possible antecedent for the null SCL in the second conjunct. This contrast pointed out in Kayne (1972) between these two French pronouns is illustrated below:

(21) a. Il_1 a dit que je l_1'ai insulté.
'He said I insulted him.'

b. *On_1 a dit que je l_1'ai insulté.
'One said I insulted him.'

c. [on_1 VP] and [e_1 VP]

These differences may not be fully conclusive: we can plausibly argue that the anaphoric pronoun in the second conjunct always receives a specific interpretation. We can expect non-specific interpretation *on* not to be able to act as antecedent for it. If *on* is interpreted as *nous* (we) - see Cinque (1988) for discussion - as it may, then it may corefer with a pronoun (*oni a dit que Jean nousi a insulté (we said that John insulted us*)) without any relevant changes to its syntactic properties. Rizzi's point concerning (21c) would remain if there were principled reasons why the null pronoun could not be indefinite. However, the same question arises for the impossibility of (5b): **Jean le présente a Pierre et envoie à Paris*. Ruling out a VP type conjunction *...le [[présente à Pierre] et [envoie à Paris]]* would follow from the Coordinate Structure Constraint. Ruling out *...[le_i présente à Pierre] et [[e]$_i$ envoie à Paris]]* could not be attributed to the impossibility for *le* to act as antecedent (recall *le* is assumed to be a major category similar to that of French *il*).

I would like to suggest an analytically more uniform analysis for this array of data. We know that the syntactic realization of a particular property as head is subject to a degree of lexical arbitrariness cross-linguistically. Thus the same function can be realized as a covert head or an overt head (e.g. French zero Q morpheme vs. Vata *la* Q-morpheme...). If overt, it may be realized as an autonomous word, or as a bound morpheme (cf. future of present in English *will* vs. French *-er*).[9] Taking this into account, we can construct an account of these differences that does not appeal to categorial differences but to an independently motivated diacritic, namely [+/-syntactically bound], i.e. [+/-clitic]. This parameter in the SCL systems is of a theoretically acceptable kind since it falls within the range of attested lexical variation. On this approach all subject clitics are consistently analyzed as head of NomP. The difference in behavior is due to the fact that French subject clitics like *il* do not have to be syntactic clitics at all, while Trentino *la* (see Safir (1985) for discussion) or French *on* are, both fairly uncontroversial assumptions.[10] The possibilities for conjunction discussed in (19) follow (or rather they follow the same way they do for object clitics - see (5), however that may be precisely): the verbal host of a clitic cannot be conjoined independently of this clitic. This conclusion is corroborated for French by differential judgments of the following sort between (22a) and (22c):[11]

(22) a. ?* Il, a mon sens, a besoin d'une bouffée d'air.
'He, in my opinion, needs fresh air.'

b. Elle, a mon sens, a besoin d'une bouffée d'air.
'She, in my opinion, needs fresh air.'

c. ** Jean le, a mon sens, respecte profondément .
John it, in my opinion, respects deeply
'John in my opinion respects it deeply.'

We mentioned earlier that subject "pronouns" can be further subdivided between exclusively nominative pronouns like *il, on, je* and others like *elle, nous*. As the pair (22a,b) illustrate, this distinction seems reflected in their clitic behavior. This second class of SCL are indeed clitics in some constructions at least (e.g. SCI or Cl), and like other SCLs disallow modification. However, except for these constructions, they may be stressed. We can accommodate their properties by stipulating that unlike exclusively Nominative SCL, they do not <u>require</u> phonological cliticization (although phonological cliticization is certainly preferred), because they are ambiguous between head of NomP and Head of DPs. As head of NomP, they are phonological clitics and belong to the weak nominative paradigm *je, tu, il, elle, nous, vous, ils, elles, on, ce*. As head of DPs, they are not clitics at all and may be stressed, i.e. they belong to the paradigm of strong pronominal forms *moi, toi, lui, elle, nous, vous, eux, elles, ca* which are not clitics. This structural ambiguity would explain why non-exclusively nominative pronouns are optional phonological clitics.[12] We end up with a three way distinction: exclusively nominative SCL are always phonological clitics. Some of them - *on* and *ce*, which we have not discussed -must also be syntactic clitics. Finally, some of them do not have to be syntactic clitics.

5.3. SCLs, Paradigmatic Regularity and Null Subjects

A third difference concerns the compared regularity of the clitic paradigms in French and in the Northern Italian dialects. Rizzi (1986) citing work by Renzi and Vanelli (1982) on 27 northern Italian dialects with "uncontroversial subject clitics", reports that in 21 of those, some clitic form is lacking for at least one element of the pronominal paradigm and even in the last 6, native speakers "often seem to consider the specification of 1st sg and 1st pl and 2nd pl clitics as a matter of strong preference rather than one of strict obligatoriness". This situation contrasts markedly with the standard French situation in which the presence of a subject "clitic" if possible at all, is obligatory. Rizzi, correctly I believe, relates this observation to the difference between Standard Italian and French that allows the first, but not the second to have silent subjects without any other morphological indication of their presence than the subject-verb agreement on the verb: "a gap in the clitic paradigm would amount to an S-structure representation with a null subject position in a non Null Subject Language..." This difference then, can be attributed to the usual difference between French and Italian regarding silent subjects without any implication regarding the categorial status of SCL. We take Standard Italian or the Northern dialects to be Silent Subject languages in the following sense: nothing beyond the content of AGRs - i.e. the agreement properties showing up on the verb[13] - is required to recover these Silent Subjects. SCL are in effect superfluous from this point of view. The phonetic gaps in the SCL paradigm do not interfere. French subject/verb agreement, on the other hand, is not "strong" enough: Silent subjects are disallowed unless recoverable in some other way. It is not that silent subjects are disallowed in French. Their licensing conditions are different. Indeed, our analysis of French SCLs makes French a Silent Subject Language as well (see Bouchard (1982), Safir (1985), Sportiche (1983)): as we have argued, in the presence of an SCL in Nom, the DP is [Spec,NomP] is silent and *pro*, in other words, a standard silent subject. Unlike in Italian, the SCL is phonetically required to license the appearance of this *pro*. Pushing one step further, we may adopt Safir's (1985) proposal that Italian (Spanish, etc..) and French differ precisely in that Italian SCLs may be silent (possibly due to the strength of AGRs), so that from the present perspective, all these languages have a full set of SCLs (i.e. NomP projections). They only differ as to whether or not the head of NomP is or must be phonetically realized, again an acceptable type of parametric variation.

6. POSTVERBAL NOMINATIVE CLITICS: SCI

We now turn to post-verbal SCL. They occur in SCI and Cl. Ever since Kayne (1972), it has been clear that Cl is a supercase of SCI and we thus begin with SCI.

6.1. General Properties of SCI

There are three properties of SCI that need to be accounted for:

(i) the position of the SCL
(ii) the root character of the construction:
(iii) the triggered character of the construction

Property (i) and (ii) are illustrated below:

(23) a. Mange-t-il?
eats-t-he
'Does he eat?'

b. *Je me demande si mange-t-il.
'I wonder if does he eat.'

Property (iii) encodes the fact that either there is a particular overt trigger for SCL or SCL triggers a particular interpretation, which we may also describe as involving a covert trigger with particular interpretive properties. The first example below illustrates the trigger case with *peut-être/possibly*, a class of adverbials that allow SCI. The other examples illustrate the second kind of interpretive trigger (respectively hortative, conditional and Yes-No questions):

(24) a. Peut être mange-t-il de la soupe.
'Maybe he eats soup.'

b. Puisse-t-il résoudre le problème à temps.
'May he solve the problem on time.'

c. Aurait-il résolu le problème à temps, nous aurions...
'Would he have solved the problem on time, we would...'

d. A-t-il résolu le problème è temps?
'Has he solved the problem on time?'

6.2. The standard account

If we adopt the standard account of the second property, as have Kayne (1984) or Rizzi and Roberts (1989) which elaborates on Den Besten's proposal for the treatment of V2 in Germanic, a root phenomenon par excellence, in terms of movement of I to C, we derive the correlation between the impossibility of SCI and V2 and a filled C, i.e. the root character of the phenomenon.

(25) a. Peut être qu'il mange (/ *que mange-t-il) de la soupe.
b. Qu' il puisse (/ *que puisse-t-il) résoudre le problème à temps.
c. Si il avait (/ *si aurait-il) résolu le problème à temps, nous aurions...
d. Je me demande si il a (/*si a-t-il) resolu le problème à temps.

Accordingly, the highest verb of the clause, say V, will have to raise to C. In order to do so and given our structural analysis, V will have to raise to I (or T or AGRs) and then to Nom in order to reach C without violating the Head movement constraint. The step through Nom will cause the head of NomP, the SCL, and the V to form an incorporated structure.

6.3. Clitic-host proximity problems and the syntax morphology interaction

6.3.1. The proximity problem

This analysis, although quite appealing, is not fully satisfactory. Consider the following differences (cf. (5c)):[14]

(26) a. Il ou elle connait bien le problèmc.
'He or shc knows well the problem.'

b. (?)Jean le ou les connait bien.
'Jean knows him or them well.'

c. *Mange-t-il ou (t-)elle?
'Eats he or she?'

We can readily handle the difference between (26a) and (26c): a preverbal SCL like *il* is not a syntactic clitic according to us: we can describe this fact by having V not raise to Nom. By contrast, the post-verbal SCL in (26c) is syntactically cliticized due to the intermediate movement of the Verb through Nom. However, the contrast between (26b) and (26c) is unexplained:

(27) (i) If both object clitics and post-verbal clitics in SCI are cliticized, why are some preverbal and other post-verbal, i.e. what accounts for the difference in positions?
(ii) Why is it that post-verbal SCL appear more strongly connected to the verb than preverbal object clitics (an observation with substantial cross Romance validity)?

Kayne (1991) offers a proposal potentially bearing on the first problem. He suggests that clitic-verb sequences are to be analyzed essentially as $[_F CL\ [_F V]]$, while verb-clitic sequences involve the movement of the verb (adjunction to $\overline{F}$) around a functional projection F to which the clitic is adjoined: $[_{F'} V\ [_{\overline{F}}\ [[_F CL\ [_F e]]$. Taking the position of the clitic as fixed in the clausal structure, Kayne (1991) proposes that a verb to the right of a clitic is actually higher in the tree than when it is to its left, ceteris paribus, a valid observation he shows, in some of the cases he discusses. Ignoring the theoretical problem of allowing adjunction of a head to a non-head (which could be fixed by assuming that the verb actually raises to a higher head position) it is hard to see how this mechanism (adjunction to $\overline{F}$) accounts for the closer proximity we are observing. If anything, this mechanism would lead us to expect the reverse given the greater "distance" between the morphemes involved. Postulating a further incorporation of the clitic to the right of the proposed V does not answer the question either, since nothing would account for why incorporation of a clitic to the left of its host would be of a looser kind than incorporation to the right of its host?

6.3.2. Syntax Morphology Interaction

How then should this closer proximity be handled? Recall Chomsky's (1992) proposal concerning affix hopping in English. In English, it seems clear that the morphological material licensed in T or AGRs may appear on the verb even though the verb does not overtly raise out of the VP to T or AGR (unlike what happens in say, French). Consequently, the traditional analysis (in fact Chomsky, 1957) postulates a process of affix lowering to V. The problem is that this lowering type of head movement creates an illicit configuration (the unbound trace of the lowered affix) which now must be rectified by LF by raising the verbal complex back to T (and AGRs). Chomsky (1992) proposes a simpler and more principled approach to this problem by treating morphological concatenation independently of the process of checking the form and ordering of affixes (I simplify his proposal here). He suggests that the verb is drawn from the lexicon fully inflected, say [[V+T]+AGRs], and that the verbal complex must raise by LF to T and AGRs in order to check (or cancel out) the properties of the inflectional affixes. English can then be seen as a language which delays raising of V to LF (i.e. raises covertly: it is a procrastinating language) while French raises the verb overtly. Let us adopt for purposes of exposition the more precise machinery postulated by Chomsky (nothing here turns on precisely what the mechanism is): a morphologically complex form [[A+aff1]+aff2]... is a pair ([A+aff1+aff2],.., (aff1,aff2,...)). The string of affixes aff_n must be canceled or checked in a left to right order (aff1 first, then aff2..) under head movement up the tree of this complex form. One immediate advantage of this approach is the treatment of suppletion, i.e. of forms such as *eat* + past = *ate*. In a system in which concatenation of morphemes is syntactic, the syntax must be allowed to check back into the lexicon after having performed the concatenation. This is a dubious property of the system since there does not seem to be any syntactic property dependent on whether or not checking of suppletive forms in the lexicon has indeed taken place: suppletive forms have exactly the same syntax as non-suppletive forms. This intermediate recourse to lexical information must therefore be eliminated, as Chomsky's proposal implies. Note as a consequence of this proposal that we can now hold that there are no lowering rules at all in syntax, surely a desirable result.

This is not to say that morpheme concatenation is always morphological rather than syntactic. We in fact expect to find both, but with different properties:

Syntactic incorporation (SI):the concatenation of morphemes is the result of the application of head movement. Because it takes place in the syntactic component as a case of a (head-)movement rule, we expect this kind of concatenation to (i) always involve upward movement, never lowering and (ii) the resulting morphology to be strictly compositional (since input from the lexicon is unavailable).

Morphological incorporation (MI): the concatenation of morphemes is a morphological (or lexical) operation. In this case, we expect to find (i) apparent cases of lowering (since a word may be generated with an affix whose licensing position is higher in the tree) and (ii) non-compositionality of the concatenation (lexical exception, i.e. suppletion, etc...).

6.3.3. Typology of cliticization

The proposal I would like to entertain capitalizes on the difference between MI and SI in order to explain the properties in (27). I would like to suggest that Cliticization (i.e. the concatenation itself, not the licensing of particular morphemes which, if required, as e.g. for inflectional morphemes, always involves syntactic movement upward to some licensing position) is either morphological or syntactic. Syntactic cliticization instantiates the way cliticization is usually thought of. It involves head movement of the clitic or the host upward, is morphologically strictly compositional and, in French and some other Romance languages, derives configurations in which the clitic is to the left of its host: in particular, this is the case of preverbal object clitics.

Morphological cliticization is a morphological operation (subject of course to syntactic checking under head movement). The clitic in this case is affixed insertion. As discussed above, it may involve apparent cases of lowering, and we should observe morphological compositionality violations. This is the case, I would like to suggest, of post-verbal SCLs. Return now to properties (27): in French, because inflectional affxation is suffixal, we expect morphologically affixed clitics to follow their host and to be as closely integrated to their host as inflectional affixes, since we are in effect dealing with one word. On the other hand, with syntactic concatenation, there is a priori neither ordering expectation (it turns out to be mostly -perhaps exclusively - prefixal in French) nor one word formed: we are dealing with two words, one adjoined to the other e.g. $[_V CL\ [_V V]]$, much the way compounds are formed.[15] Furthermore, as expected we find compositionality violations with post-verbal SCLs, not with preverbal clitics. In French such a violation is illustrated by the epenthetic /t/:[16]

(28) a. Il a téléphone / a-**t**-il téléphone.
'He has called/ has he called.'

b. Elle aura fini / Aura-**t**-elle fini.
'She will have finished / will-have she finished.'

The same conclusion extends to the Northern Italian dialects such as Trentino or Fiorentino discussed earlier. Recall that in these languages, preverbal SCL unlike French SCL are indeed clitics syntactically. They also allow SCI constructions (examples from Brandi and Cordin (1989)):

(29) a. Quand es-tu arrivé?
b. Quando se'-tu arrivato? (Fiorentino)
c. Quando set riva? (Trentino)
'When did you arrive?'

Not only do post-verbal clitics exhibit allomorphy (or morphologically conditioned epenthesis):

(30)

Preverbal SCL		post-verbal SCL	
IIs	tu	tu	Fiorentino
	te	t	Trentino
IIIsm	e/gli	egli	Fiorentino
	el/l	lo	Trentino
IIIsf	la	ella	Fiorentino
	la	la	Trentino

but Fiorentino exhibits cases in which the post-verbal SCL actually occurs between the verbal root and AGRs:

(31) a. Icche **gl'** han**no** fatto?
b. Icche ha(e)-**gli-no** fatto?
'What have they done?'

As Brandi and Cordin note, examples such as (31b) are not expected if the ordering of the morphemes resulting from the morphological merger is entirely determined by the linear order of syntactic categories in a tree. In effect, we even find cases in violation of the Mirror Principle if what was involved was concatenation under head movement. Under an MI analysis, there may be (although it certainly appears to be exceptional) cases of mismatch between linear ordering of morphemes and ordering of the syntactic projection in which they are checked because the two are dissociated: there may be exceptional complex forms ([A+aff2+aff1]..,(aff1,aff2,..)). We would treat this as a case of complex allomorphy (hence exceptional, but allowed in principle) swapping the order of affixes in MI.[17]

6.3.4. SCI and V movement

Because of the existence of the phenomenon called affix lowering, we know that the presence of a particular morpheme on some head says nothing of the syntactic position of that head in the overt syntax. Applied to the case of SCI, the fact that Nom - the SCL - appears adjoined to the right of the verb says nothing about whether verb raising has applied or not. The fact that SCI only occurs in possible V2 contexts makes overt raising to C a plausible analysis. We now turn to CI. As we will see, the properties of CI challenge this conclusion and I will propose an analysis of the root character of SCI or CI not involving overt V to C.

7. POSTVERBAL NOMINATIVE CLITICS: CI

Recall the basic examples and the basic problems raised by CI:

(1) a. Jean est-il malade.
b. Depuis quand Jean est-il malade.

Just like SCI, complex inversion is a root phenomenon; It is consequently reasonable to suppose that it involves raising of I to C, the classic V2 effect. Given that Wh-movement is to [Spec,CP] and the subject DP intervenes between the wh-phrase and the highest verb (as in (1b)), where is this DP subject? This is what Rizzi and Roberts call the "landing site problem". Finally, why may a pronominal copy of the subject cooccur with a full DP subject, making complex inversion a construction with two subjects. In the words of Rizzi and Roberts, the existence of two apparent subjects in CI pose the Case problem - how is each subject assigned Case - and the source problem - where do these subjects originate.

7.1. The landing site problem: I to C and root phenomena

7.1.1. Some remarks on the availability of I to C

The landing site problem arises primarily because of the assumption that CI involves I to C in the overt syntax. It is tempting to drop this assumption, as is suggested in Morin (1979) or Noonan (1989; 1992), provided that a plausible account of the root property can be constructed. If there was no overt I to C, there would not be such problem: the tensed verb could either be in AGRs with the subject in [Spec,AGRs] or [Spec,NomP] or have raised to Nom with the subject in [Spec,NomP]. The only obstacle to assuming that there is no I to C in CI, it seems, is the account of the root character of CI. Let us examine it in more detail. Initial accounts of CI (and V2 and related phenomena) built on the observed correlation between Cl and the presence of overt complementizer. Postulating I to C accounts for the correlation because an overt complementizer fills the C slot, making it unavailable for a raised I. Surely, however, as Rizzi and Roberts discuss, this account is insufficient, and at the very least needs a condition along the lines of (i) below to supplement (ii).[18]

(32) i. The CP is not lexically selected (restricting I to C to main clauses, adjuncts).
ii. The C position is not filled.

The necessity of some additional condition with the effect of (32i) is supported by the instances Rizzi and Roberts point out of selected "complementizer-less" clauses which do not allow I to C even if the language has I to C: infinitive clauses on a widespread scale, *that*-less tensed complement clauses in English. Even if the cases of silent Cs could somehow be dealt with, there are also cases of overt clitic complementizer in Hebrew or Irish in which a clitic C lowers to I (see Shlonsky (1988), McCloskey (1992)), or very possibly French "interrogative" *que* (as in e.g. *que fais-tu? / What do you do* in which *que* and *fais* are cliticized to each other) that do not show the complex of properties associated with I to C. The necessity of (32ii) is supported, according to Rizzi and Roberts, by cases of unselected clauses with filled Cs disallowing CI:

(33) a. Peut-être Jean est-il parti/ Peut-être que Jean est(*-il) parti.
'Maybe John is gone / maybe that...'

b. Aurait-il mangé/ Si il avait mangé (*aurait-il...).
'Would he have eaten/ If he had eaten (*had he eaten).'

Some questions arise however. The first clause appears somewhat too strong: certain varieties of English, for example, allow I to C in embedded interrogatives (as e.g. *John wonders should he go to the store*). The second clause, if it is indeed a primitive, is suspicious; it seems to condition the availability of a syntactic process (I to C) on a phonological one ("filled" or not); it introduces a degree of arbitrariness (if it is a primitive, the distribution of unfilled Cs is arbitrary); most important however is the following difficulty: French CI or English I to C is impossible in statements in main clauses lacking an overt complementizer. This strongly suggest that I to C is a triggered rule. It applies only if some specific property (question, doubt...) triggers it. This points in the following general direction that I will not elaborate on in detail here (but see Sportiche (1995b)): I to C (or more broadly head movement) is fundamentally a triggered process. C sometimes contains a property, say, [+Q], that must be realized. In selected environments, this licensing does not require the raising of a lower I to C because insertion of a subordinator - e.g. the complementizer - typically (but not always, cf. the varieties of English alluded to above) is available.[19] Although in principle, this licensing could arise from raising of a lower I to this C, this is prohibited by Chomsky's (1992) Principle of Greed: movement is motivated only to license a property of the moved item. Here I has no property of itself to license.

If the [+Q] property appears on an unselected C, the insertion of a subordinator may be excluded (in main clauses, by definition). The only way to licitly license this property is to raise I to C. This means that we must interpret the Principle of Greed as an Economy Principle and not as an absolute prohibition.[20]

This type of account is probably rendered necessary as soon as we adopt Chomsky's view of the morphology/syntax interaction described earlier. It would indeed be extremely surprising not to find cases of covert V2 - with the same distribution of properties as overt V2 - analogous to cases of covert V to T raising analogous to French overt V to T raising. But more generally, the fundamentally identical properties (in terms of locality, etc..) of cases of head raising (V to T) and head lowering (affix lowering) demands treatments of head movement that are neutral between raising and lowering - i.e. between overt and covert raising. This naturally applies to V2 effects.

Consequently, and most importantly for our purposes, there is no variation in the configurations allowing and thus requiring I to C. As usual, this raising may take place either in the overt syntax or in the covert syntax, depending on the language and the particular construction. But this is the only type of variation allowed.

7.1.2. Covert I to C in French

Nothing then really turns on whether I to C movement is overt movement or covert movement. We have to make sure that a complementizer that counts as filled at S-structure also counts as filled at LF. If this is insured, in principle the cases of permissible overt I to C and covert I to C are going to be identical. This can be achieved by the null assumption: if a complementizer position counts as "filled" at S-structure, it means that its licensing does not require I to C, hence does not require I to C at LF. Consequently, we are free to assume that CI involves in fact covert I to C without affecting the account of the overall distribution of CI. This is what I am going to assume now. We should expect never to find really free alternations between I to C and lack thereof. Let us now return to the cases in (33) which look like such cases:[21]

(33) a. Peut-être Jean est-il parti/ Peut-être que Jean est(*-il) parti.
'Maybe John is gone / may be that...'

b. Aurait-il mangé?/ Si il avait mangé (*aurait-il...).
'Would he have eaten/ If he had eaten (*had he eaten).'

Rizzi and Roberts assume that the *que* headed clause in (33a) is unselected.[22] It seems at least as plausible, extrapolating from Morin's (1985) work on *voilà* clauses, that the *que* clause in *peut-être que* is actually lexically selected as a complement by the higher adverbial predicate *peut-être..,* a reasonable assumption given the semantics of these adverbials, which is reminiscent of the corresponding adjectives *possible, probable, certain, difficile, rare* which do take a clausal argument. The independence of the two structures is supported by the existence of adverbs which allow I to C but not a *que* clause as e.g. à *peine (as soon as), encore (untranslatable discourse particle)* ***viz.*** *A peine est-il arrivé/*A peine qu'il est arrivé; Encore aurait-il fallu../*encore qu'il aurait fallu* and which, correlatively, are not predicates at all. Furthermore, the following lack of alternation is unexpected if indeed, all that is at stake is the filled/unfilled alternation in otherwise identical structures or the selected/non-selected distinction:

(34) a. Je crois que peut-être qu'il est parti.
'I believe that maybe that he left.'

b. *Je crois que peut-être est-il parti.

In our terms, the impossibility of SCL Inversion (which incidentally is always optional in standard French, i.e. freely overt or covert if we are right) in (33a), (34a,b) must be attributed to the Principle of Greed - i.e. to the existence of an alternative to I to C (roughly, insertion of *que*, a subordinator, which is not available in main clauses).

The cases of (33b) are different since we are dealing throughout with non-main-clauses. Here too, the structures differ: in inverted structures, the verbal mood inflection is conditional; in non-inverted structures the inflection on the verb is indicative. Thus, no real optionality is found here either.[23]

7.1.3. The distribution of -il and the Principle of Greed

I propose that there is no overt verb movement to C in Complex Inversion constructions nor of course in SCI constructions. I also propose that SCLs are licensed in the position head of NomP. In the case of post-verbal SCLs, which I assume are morphologically incorporated to the verbal complex, licensing of the SCL requires movement of the verbal complex to Nom. However, as we saw, post-verbal SCL are licensed only in questions (or other triggered environments), i.e. only in cases in which I to C is otherwise required. How then is the licensing of a post-verbal SCL prevented in other contexts. To put it differently, what excludes the raising of the verbal complex to Nom in environments lacking an appropriate trigger in C. We need a three way distinction:

1. Type 1 heads such as Tense or AGR, which may be generated on V must be able to force V-raising to T or AGR (see below) to get themselves licensed in a way consistent with the principle of Greed. Thus, V+T raises to T to license T.

2. Type 2 heads such as [+Q] Cs must be able to trigger raising of the verb to C (more precisely of I to C, as V to I is independently motivated), in violation of (the economical interpretation of) the principle of Greed, if no more economical option is available, to get themselves licensed. Thus, I raises to [+Q] C only in unselected contexts.[24]

3. Type 3 heads such as *il* which cannot trigger raising at all in order to get themselves licensed and are therefore (i) either generated where they are licensed (preverbal SCLs), or (ii) are generated on some other head H and can only be licensed if H has to raise to or through their licensing position for independent reasons (post-verbal SCLs licensed though the independent raising of I to C).

The distinction type 1/ type 2 is necessary as a consequence of Chomsky's proposal concerning morphology and lowering. The distinction between type 3 and the others, the innovation we introduce, appears necessary as well. Recall that complex forms are assumed to be generated fully inflected and that the affixes are licensed through head movement either overtly or covertly. This proposal raises the following question: how complex can complex forms get? Thus, we would in principle expect to find cases of (35a) with (35b) reading:

(35) a. I [e] [e [John left-that-believe]].
     b. I believe [that John left].

in which the complex embedded form would be properly licensed by successive movements from its V position to I to C to V to I - all possible instances of head movement. Such cases appear rare. In order to account for this restriction we in effect need a notion of "natural complex word". The distinction between type 3 heads and the two others would be of this nature: in effect a type 3 head, when attached to another head does not form a natural complex word with it; its licensing requires some exceptional rescuing device (here raising to C) which will allow the parasitic licensing of this type 3 head.

Leaving type 2 heads aside, I suggest relating the distinction between type 1 heads (AGR, ASP..) - call them X-strong heads - on the one hand, and type 3 heads (Nom) - call them X-weak heads - on the other to whether or not the property expressed by the head is lexically related to X: AGR is a Case related property, and T(enses) can lexically differ as to their Case assigning properties. AGR is hence a T-strong head forcing the raising of T+AGR to AGR (hence of V+T+AGR to AGR).[25]

Nom however is not an X-strong head for any head in the verbal system. Nom cannot trigger raising on its own of some other verb-related head to Nom.[26] Consequently, a post-verbal SCL is never licensed unless the verb raises to a position higher than Nom for independent reasons, licensing the post-verbal SCL on the way. In other words, it may licitly be affixed to a verb only in V2 triggering contexts.

7.1.4. Excursus on other varieties of French

Yes/no questions in two varieties of French can be quite naturally analyzed very much along the same lines. In each, post-verbal SCL have been replaced by an invariable suffix, namely *-tu* in varieties of Quebec French and *-ti* in varieties of popular French documented in Morin (1979):

(36) Jean/Je peux-ti ajouter quelque chose?
'Can Jean/I add something?'

Here, the post-verbal particle may cooccur with a preverbal SCL, something impossible in the standard variety. Otherwise this construction has a similar distribution and allows the same subject DPs as the standard CI or SCI. The analysis is straightforward: the post-verbal affix is a morphologically concatenated question morpheme (as Noonan (1989) suggests) that must eventually be licensed by the verbal complex raising to C at LF. Raising to C however does not take place overtly for the same reasons as in the standard variety.

7.2. The double subject problems

For Rizzi and Roberts or Kayne (1984), the post-verbal pronominal form is essentially a DP subject cliticized. Hence the Case problem, the source problem and, as an additional question, the difference between pronominal and non-pronominal DPs in regard to their potential ability to appear post-verbally. Since we do not treat these post-verbal elements as DPs at any level, it is clear that none of these problems arise.

7.2.1. The Case problem

Rizzi and Roberts or Kayne suppose that SCI or CI involve I to C. Post-verbal subjects are then expected to be possible (just as in English yes/no questions or German V2 structures). The impossibility of (37a):

(37) a. *A Jean parlé?
'Has John spoken?'

b. A-t-il parlé?
'Has he spoken?'

needs to be accounted for and in a way that allows (37b). Rizzi and Roberts propose, following Szabolcsi's (1983) suggestion, that I to C in French, contrary to English, destroys the context for Nominative Case assignment. Rizzi and Roberts postulate that Nominative Case assignment in French is only right to left: raising of I to C yields a Case Filter violation in (37a): *Jean* has no Case. In (37b), the pronominal DP subject can satisfy its Case requirement by incorporating into its Case assigner, as Baker, Johnson and Roberts (1988) have suggested. The first part of this proposal is difficult to reconcile with the current conception of (structural) Case assignment. Structural Case assignment is the reflex of a Specifier head relationship. This raises two problems: first we expect English and French to be treated exactly alike from the point of view of Case theory - which is incompatible with Rizzi and Roberts's proposal; secondly nominative and accusative Case assignment should function exactly alike. Nominative arises from a Spec/head relationship between T in AGRs and the latter's Specifier position. Accusative arises from a Spec/head relationship between V in AGR_o and the Specifier of AGR_o. If T to C destroys the configuration for nominative Case assignment, why does not V to T, which occurs systematically and is parallel to I to C, destroy the configuration for Accusative Case assignment?

(38) C DP AGRs T T DP AGR_o V

If pre or post-verbal SCL are not analyzed as DPs, this problem does not arise at all. (37a) is excluded simply because French lacks overt I to C.

Kayne (1984) conjectured that CI was possible in French only because post-verbal SCI were clitics (and more generally that a language may have CL-like constructions only if the equivalent of post-verbal SCL were clitics, thus partially explaining its rarity). Rizzi and Roberts's account has the property of theoretically substantiating this conjecture: for them, that incorporation in (37b) must take place is motivated by Case theoretic considerations: the post-verbal SCL is a DP, hence needs Case. It is not in a Case receiving configuration, hence incorporates.[27]

Kayne's conjecture is also derived within the present account. CI involves inversion of some sort by definition but not Verb/Subject. Instead, it is inversion of two normally adjacent heads H1 and H2 normally occurring in this order. If the inverse order cum adjacency is found, because of the Head movement constraint, this can only occur if one incorporates to the other in one way or another. Hence, in a sense each of them must be a clitic on the other.

7.2.2. The source problem

Clearly, the source problem, a problem of potential undergeneration, does not arise under the present proposal. In Cl constructions, there is only one subject. However, converse problems arise, i.e. problems of overgeneration: so far we have assumed that nothing in particular about the availability of the NomP projection (except that it may only occur in clauses having nominative, i.e. tensed clauses in French). Consider first the case of preverbal SCL, e.g. *il a mangé/he ate*. Here, we have the following relevant substructure:

(39) Nom AGRs T V
il

the SCL *il* is not (necessarily) a syntactic clitic. The verb does not (have to) raise all the way to Nom but must raise at least to AGRs (the standard analysis, cf. Pollock (1990)). Let us now look at the DP subject. Sportiche (1995a) shows that, in French and when it is possible to tell, a silent DP, i.e. *pro,* doubling a clitic head of ClP must raise to [Spec,ClP], at least. It is difficult to show directly that this is the case in the present situation, but it is not unreasonable to extend this conclusion here.

With a preverbal SCL, this DP cannot be overt in standard French. Provided that this DP is required to raise to [Spec,NomP] overtly,[28] surely the null hypothesis given the behavior of silent DPs, we may, as mentioned earlier, encode this impossibility in a version of the "doubly filled Comp filter" extended to the Nom system: Nom and [Spec,NomP] cannot be simultaneously filled (in the relevant sense) at S-structure.[29]

Turn now to post-verbal SCLs.

(40) Nom AGRs T V

The difference here is that both overt and silent DP subjects are allowed. The latter case raises no particular question: it raises just as in the previous case to [Spec,NomP]. But the former does: why are overt DPs suddenly allowed? By uniformity, an overt DP should be considered to have raised to [Spec,NomP] as well. Furthermore, we have good independent evidence that tensed verbs always raise to AGRs overtly (Pollock (1990)). The only conclusion compatible with the doubly filled projection approach is that the combination V-SCL raises no further than AGRs. It does not raise to Nom, hence does not create a doubly filled projection. This means that the affixation of the post-verbal SCL could not have resulted from the V raising to (or past) Nom. Rather, concatenation

V+SCL must have occurred independently of head movement of V to Nom. This is what we had concluded earlier: it is precisely the fact that post-verbal SCLs are subject to MI that allows CI to exist in French.

It is worth pointing out the logical relations between the various proposals we make at this point. The fundamental proposal we make, i.e. that SCLs are the head of Nom, and that there is no head movement to C in SCI and CI constructions is independent from the analysis of post-verbal clitics as involving MI. This proposal could be maintained even if the surface position of the post-verbal SCL was due to movement (of the verb or of the SCL). It would probably indicate that tensed verbs in French routinely raise to Nom (or that there is syntactic lowering of the SCL).

7.3. Some direct evidence for lack of overt I to C

7.3.1. Distributional Evidence

If I to C does not take place overtly in French, nothing special need be said about the position of the DP subject in CI: it occurs in [Spec,NomP]. The V+SCL combination is in AGRs. This is illustrated in (41a).

In Kayne's or Rizzi and Roberts's analysis, I to C does take place overtly: V+SCL is in C and the DP subject in some intermediate position between C and [Spec,CP], where wh-phrases occur. This is illustrated in (41b).

(41) a. $[_{CP}$ XPwh C $[_{IP}$ Nom DP AGRs T V V+SCL

b. $[_{CP}$ XP_{wh} DP V+SCL C $[_{IP}$ ...

The two analyses make quite different predictions concerning "IP" initial material: in CI or SCI, this material is predicted to occur to the right of a wh-phrase and

(i) to the left of the V+SCL complex or to the left of an overt subject if any, according to the first analysis.
(ii) to the right of the V+SCL complex, according to the second analysis.

The following examples illustrate a possible case. Consider the temporal clausal adjunct (42a). Elements of its distribution are illustrated in (42b) (the symbol *^ indicates that the adjunct clause in (42a) cannot appear in the position of ^ in (42b), a bare ^ indicates that it may):

(42) a. Quand le vote a eu lieu
when the vote took place

b. ^ les électeurs sont ??^ allés à la pêche
the voters went fishing

c. Dans queue ville ^1 (les electeurs) sont-ils ??^2 allés à la pêche
In which city (the voters) went fishing

The judgments seem preserved in (42c): although position 1 is perceived as a natural position for the intervening temporal clause, position 2, although not impossible is perceived as less natural, stylistically more marked. This observation is explained if the subject DP and the verbal complex occupy respectively the same positions as in (42b), and mysterious otherwise. The post-verbal position 2 in (42c) is particularly significant. Both Kayne (1984) and Rizzi and Roberts (1989) consider that because of I to C, the pre-C position hosting the proposed subject counts as a [Spec,IP] so that there are in effect two IPs: the original IP and a derived IP. Position #1 is thus a pre-IP position.

7.3.2. Binding Evidence

The position of the overt subject in CI must be construed the way a (possibly derived) A-position usually is (however this should be defined). This is shown by the possibility of sentences like *De qui$_2$ [le père t$_2$] est-il venu? (Of whom did the father come)*: extraction from elements proposed in A-bar position is not possible (cf. **De qui te demandes-tu quel frère est venu / De qui$_2$ [le père t$_2$] Jean l$_2$'a vu (Of whom do you wonder which brother came, Of whom the father John saw him.).* This is as expected under both Rizzi and Roberts's and the present analyses (but is the opposite conclusion of Kayne's (1984)). In the present analysis, the subject is in the normal subject position. In Rizzi and Roberts's, it is the Specifier of a derived IP (so that raising to C is in effect a structure building rule).

SCI, *est-il venu,* is treated differently by each analysis:

(43) a. C [$_{NomP}$ pro [e [$_{ARGsP}$ [est-il....
b. [$_{IP}$ e [est-[il]$_2$ [IP t_2 [...

Consider now the following sentences:

(44) a. Si Jean$_2$ était venu, il$_2$ aurait decidé...
'If John had come, he would have decided...'

b. *Il$_2$ aurait, si Jean$_2$ était venu, decidé...
'He woud have, if John had come, decided...'

c. *Aurait-il$_2$, si Jean$_2$ était venu, decidé...
'Would he, if John had come, decided....'

(44b) illustrates a Principle C effect: the pronoun *il* or the *pro* c-commands the name *Jean.* Why is this effect also found in (44c)? According to structure (43a), it is for an identical reason: the *if* clause is lower than AGRs and thus c-commanded by *pro*. According to structure (43b),the reason is less clear. The *if* clause can appear, say in an adjoined position, IP initially (an observation corroborated, within Rizzi and Roberts's proposal), by the possibility of *Quand, si Marie est malade, Jean pourra-t-il lui rendre visite (When, if Mary is sick, will John be able to visit her)*. Adjoined to the lower IP in (43b), it should not trigger any principle C effect as it is not within the c-command domain of any coindexed DP. There is also no reason to count the incorporated head it of the subject DP as relevant for the binding theory.[30]

8. SUBJECT EXTRACTION AND THE ROLE OF NOMP

There are two interrelated questions that we have not addressed yet. The first is the function of the NomP projection. The second concerns the deviance of simultaneous subject extraction and SCI illustrated in (45):

(45) Qui$_2$' a-t-il$_2$ mangé de la soupe?
who has-he eaten soup

The deviance of (45) is actually not fully representative of the data. Thus, as noted in Grevisse (1980), subject extraction and SCI, i.e. Wh-movement of the subject in CI constructions, are compatible for example in questions and in exclamatives:[31]

(46) a. [Combien d'entre nous]$_2$ t_2 auraient-ils$_2$ droit.
'How many among us would-have-they the right.'

b. [Quel féroce magicien]$_2$ t_2 a-t-il$_2$ enfermé ton secret?
'Which fierce magician has-he locked up your secret?'

Specificity of the subject seems to be a determining factor allowing cooccurence of SCI and subject extraction. The correct notion, say specificity, however it should be characterized (see Enç (1991) and references therein), is strongly reminiscent of that relevant for accusative clitic doubling (in River Plate Spanish, see Jaeggli (1982), Porteno Spanish, see Suñer's work, e.g. Suñer (1991), Rumanian, etc.) or "accusative case marking" in Turkish (see, Enç (1991)). Sportiche (1995a) suggests that the function of AccP, the Accusative counterpart of NomP whose head hosts accusative clitics, is precisely that of a "specificity" licenser. Extending this idea to NomP not only provides an answer to the first question, but explains the difference between (45) and (46). The examples in (46) are acceptable precisely because it is natural to interpret the variable bound by the wh-quantifier as ranging over a set previously introduced in the discourse. After all, this conclusion is not surprising as the cases under discussion are cases of Clitic Doubling involving Nominative instead of Accusative.

If this approach is correct, the specificity requirement imposed by the presence of the SCL, i.e. of NomP, should be observed in non wh-structures as well. Non-wh-DPs incompatible with a specific interpretation should not be able to occur as subjects in CI constructions. This prediction is difficult to test if specificity is correctly characterized in terms of prior introduction of some (restricted) set of possible referents for it might rely not so much on the internal structure of DPs as on the discourse conditions of use of DPs (e.g. why couldn't *qui* in (45) be interpreted as specific in particular discourse conditions). Fortunately, there is a class of DPs which cannot receive a specific interpretation, namely indefinite DPs modified by a relative clause in the subjunctive mood (they cannot appear with a definite determiner). These DPs cannot appear in the CI construction. Indefinite DPs modified by a relative clause in the indicative mood are ambiguous between a specific and a non specific interpretation. They may appear in CI constructions. When they do, however, they can only be interpreted as specific:

(47) a. Je cherche un homme qui sache/sait cuisiner.
I am looking for a man who could-subj / can-ind how to cook
(specific reading only / ambiguous)

b. Un homme qui *sache/sait cuisiner chercherait-il un tel poste.
a man who could-subj/can-ind cook would-seek-he such a position
'Would a man who can cook seek such a position?'
(specific reading only).

Given the similar treatment we propose for French and Northern Italian SCL, we expect that similar conclusions should hold of the Northern Italian dialects. Suñer (1992) suggests that they do despite prima facie evidence to the contrary, such as the possibility of doubling quantifiers such as *nobody* with SCLs. She points out that in Porteno Spanish, doubling of a direct object *ninguno (nobody)* is possible and suggests that this negative quantifier in all these cases is interpreted as partitive.[32] Whether this is correct is unclear. As I mentioned earlier, specificity as I have been using it may not be the right notion (see Sportiche (to appear) for a different suggestion). It seems that the right set of DPs should at least include specific DPs. The crucial factor here is that we find parallel behavior in subject doubling and object doubling.

9. CONCLUDING REMARKS

The analysis of subject clitics we advocate for French and, more generally, Romance, is in line with that suggested in Sportiche (1995a) for object clitics. It also marks a return to one of the earliest analysis of SCLs, namely that of Kayne's (1972). Kayne suggested that NPs be generated in the form [NP -SCL] (e.g. *[Jean -il]*, or *[lui -il]*), where either the strong pronoun lot or the SCL it be deleted, except in special contexts such as CI. The idea that the SCL and the DP are distinct entities is the fundamental idea we have pursued. Essentially, we have departed from Kayne's proposal only in requiring that the configuration DP-SCL is not base generated as such but has to hold at LF.

NOTES

[*] This article is dedicated to the memory of Osvaldo Jaeggli. Its content, I hope, is no mere "bourdonnement de mouches à ses oreilles".

[1] Although this is not necessary for the conceptual basis of my proposals. However, it seems quite plausible for subject clitics as we will see later, and for object clitics for the reasons discussed in Sportiche (1995a). Alternatively, in certain cases, the clitic heads the XP (XP* or the head of the Specifier of XP*, see Koopman (1993)) referred to below and the head H^{O} is silent. Correlated changes to the text below would be necessary if this alternative were correct.

[2] Syntactic clitics cannot be separated from their host by any syntactic process and are necessarily phonological clitics. Phonological clitics are syntactically autonomous if they are not syntactic clitics as well.

[3] On *on, ce sa* Subject Clitics which we do not discuss here, see Kayne (1972), Jaeggli, (1982) and Rizzi (1986).

[4] The difference with the analyses mentioned has to do with the modernized treatment of subject/verb agreement that we adopt here.

[5] One immediate and somewhat surprising consequence of this, which we do not discuss here, is the apparent position of the subject: the subject appears not to be in what is usually deemed to be the normal subject position, i.e. [Spec,AGRs]. Rather, it appears higher in the structure in [Spec,NomP].

[6] (13c) is not ungrammatical throughout. Speakers who routinely accept (13a), as e.g. certain Quebec French speakers, also allow a quantifier. The basic point remains however for Standard French (although maybe not for other varieties).

[7] Also known as the doubly filled Comp filter. Of course, why the doubly filled CP filter exists needs to be explained. The idea of competing for one particular entity could be preserved if the competition is for a particular property instead of a position.

[8] Or more precisely *[pra$_2$ tous t$_2$]*, given *[eux tous]* and **[tous eux]*, cf. Sportiche, (1995a) for discussion.

[9] I will discuss later a further subdivision between lexically bound morphemes - regular inflectional morphemes - and syntactically bound morphemes - i.e. morphemes which become bound as a result of a syntactic operation of head movement and are therefore not as closely integrated in their host as the others: these I would call clitics.

[10] Note that Rizzi's (1986) discussion of reordering phenomena between SCL and negatives clitics in northern Italian only shows that SCLs belong to a different clitic cluster than object clitics. It does not show that they are not syntactically cliticized.

[11] Kayne (1984) attributes to Hirschbühler the report that some dialects accept (22a) as well formed. Another contrast is noted in Kayne (1984): *Il sont partis. Qui, ils? (They left. Who, they?)*; **Pierre les a vus. Qui, les? (Pierre them saw. Who, them?)*. *Les* obligatorily being a clitic, it may not appear stranded. On the other hand, *ils* is not. There is more to say about this topic, suggesting further distinctions: *on* behaves like a syntactic clitic as suggested by the data in (6) but allows *on a mange mon gateau. Qui, on? (Someone ate my cake. Who, someone?)*; the ungrammaticality of (6b) with on is less sharp than that of of (5b). This suggests a more refined distinction amongst syntactic clitics, making SCLs not as closely tied to their verbal host as OCLs.

[12] Left open is the deviance of Cl *Elle est-elle partie (She is-she gone)* under normal stress and intonation, as compared to the acceptable *Marie et elle sont-elles parties (Marie and her are they gone)*.

[13] We differ here from Rizzi's understanding, who takes the SCL to be the realization of subject agreement.

[14] These judgments are I believe duplicated although all shifted towards the worse with *et (and)* instead of *ou (or)*.

[15] This distinction has sometimes been implemented differently by distinguishing two types of head movement: substitution (into a subcategorized slot) and adjunction, which would correspond to the MI/SI distinction. In effect, we are claiming that head movement as substitution is morphological incorporation subject to syntactic checking under head movement. The converse seems false however. There are, if we are right, cases of MI not involving subcategorized slots as locus of substitution in any reasonable sense: post-verbal SCL for example.

[16] Its appearance is morpho-phonologically conditioned (between a verb with appropriate phonological ending and a SCL cliticized to it beginning with a vowel). It cannot be analyzed as third person agreement being sometimes deleted, as it never shows up in any other context where it should, e.g. in liaison contexts.

[17] I will not discuss other cases of post-verbal clitics, i.e. object post-verbal clitics found in positive imperatives: *Mange le, Donne le lui (eat it, give it to him)*. Extending the present proposal to these cases is suggested by the relative lack of parallelism between what is found in preverbal position and what is found in post-verbal position: *Jean m 'en donne, Donne m 'en; *Jean moi (z-)en donne, donne moi z-en* (Note the epenthetic z- intervening in the clitic string); **Jean en moi donne, donne z'en moi; (John to me of it gives, give (some) of it to me)*, or *Pierre l'y déposera, ?* dépose l'y (John will drop him there, drop him there)*.

[18] They also suggest to derive the first condition from the Projection Principle. See Sportiche (1995b) for further discussion.

[19] Selection also must be licensed: Sportiche (1995b), expanding on Koopman (1994), suggests that this is done by incorporating this C into its selector.

[20] This approach naturally extends to other cases. I would interpret Negative Inversion as in *(Mary said that) under no circumstances would he agree to that* exactly as a case of licensing by I to Neg of the head Neg of NegP, whose Specifier position is occupied by the negative adverbial phrase, i.e. exactly as cases of unselected C discussed below.

[21] Rizzi and Roberts also cite other types of examples which raise the same kind of questions or are less convincing for a variety of reasons: for example the popular French pairs: (i) *qui que tu as vu/qui qu'as-tu vu (who (that) did you see)*. Besides the questions of the text, an additional problem here is that the register allowing doubly filled comp simply lacks the constructions SCI nor CI -even, of course, in contexts lacking overt complementizers.

[22] They also assume that the small class of items *(peut-etre, probablement, sans doute, difficilement, rarement..)* allowing these alternations appear in [Spec,CP]. This seems incompatible with the acceptability of Adjunct extraction like: *Pourquoi as-tu dit que peut-être que Jean a démissionné t (why did you say that may be John resigned)*. If they did, this should be on a par with adjunct extraction out of wh-islands.

[23] *Si* clauses and their inverted counterpart differ in interpretation as well, I believe. The former may be interpreted either as (past) hypothetical or (present) counterfactual, the latter only as counterfactual.

[24] It is clear that the type 1/type 2 distinction is for expository purposes only here. It is an accidental fact where a particular head is generated. What matters is that it transits through or ends up in its licensing position by LF.

[25] This might be extended to type 2 heads as well with an extended notion of lexical property: [+Q] would be an I-strong head because of the kind of selectional restrictions between Cs and T or T+AGRs.

[26] If anything, the function of Nom is to license specificity of the associated XP*, hence could only license the raising of the head of XP* to Nom, see section 9.

[27] Although the link between Case and Incorporation seems to me theoretically highly desirable, the motivation that has been given for it seems to me unconvincing. For example, English passive *-en*, which motivates Baker, Johnson and Roberts's (1989) analysis leading to the view of the relation Case/Incorporation can, I believe, be reanalyzed as heading a projection, the passive Voice, in a way very similar to my current analysis of clitics heading Clitic Projections, and not as a nominal at all (see Sportiche (1995b) for further discussion).

[28] In Chomsky's (1992) terms, the property licensed on this DP by the raising, to which we return below, is strong, and must thus be licensed prior to spell out.

[29] Not all varieties of French exhibit this prohibition. Thus Quebec French has been argued to allow an overt - non-dislocated - subject DP doubling a SCL (see Bouchard, (1982), for example). Interestingly, Quebec French also appears to lack the doubly filled Comp filter.

[30] See Sportiche (1995a) for discussion of comparable cases with object clitics.

[31] It should be remembered that CI belongs to a fairly formal register. Intuitions in complex cases are not as secure as in more spoken varieties of French. This caveat given, I may add that, although the wh-question examples given below are acceptable, such cases are most natural with exclamatives.

[32] My judgments on French are not straightforward. Thus, while *personne* in simple CI is impossible *(*Personne n'a-t-il mangé/Nobody has-he eaten = Has anybody eaten)*, it improves in *why* questions *(Pourquoi personne n'a-t-il mangé/why hasn't anybody eaten)*.

REFERENCES

Baker, M., K. Johnson and I. Roberts: 1989, 'Passive Arguments Raised', *Linguistic Inquiry* 20, 219-251.

Brandi I. and P. Cordin: 1989, 'Two Italian Dialects and the Null Subject Parameter', In O. Jaeggli and K. Safir, eds., *Null Subject Parameter*, Kluwer Academic Publishers.

Chomsky, N.: 1957, *Syntactic Structures*, Mouton, The Hague.

Bouchard, D.: 1982, *On the Content of Empty Categories*, Doctoral Dissertation, MIT.

Chomsky, N.: 1992, 'A Minimalist Program for Linguistic Theory," *MIT Occasional Papers in Linguistics* 1, Cambridge, Mass.

Cinque, G.: 1988, 'On si Constructions and the Theory of Arb', *Linguistic Inquiry* 19, 521-581.

Enç, M.: 1991, 'The Semantics of Specificity', *Linguistic Inquiry* 20, 1-25.

Grevisse, M.: 1980, *Le Bon Usage*, 11ème edition, Editions Duculot, Paris-Gembloux.

Jaeggli, O.: 1982 *Topics in Romance Syntax*, Foris Publications, Dordrecht.

Jaeggli, O. and C. Silva Corvalan: 1986, *Studies in Romance Linguistics*, Foris, Dordrecht.

Jaeggli, O. and K. Safir: 1989, *The Null Subject Parameter*, Kluwer Academic Publishers.

Kayne, R.: 1972, 'Subject Inversion in French Interrogatives', In J. Casagrande and B. Saciuk, eds., *Generative Studies in Romance Languages*, Newbury House, Rowley.

Kayne, R.: 1975, *French Syntax: The Transformational Cycle*, MIT Press, Cambridge, Mass.

Kayne, R.: 1984, *Connectedness and Binary Branching*, Foris Publications, Dordrecht.

Kayne, R.: 1991, 'Romance Clitics, Verb Movement and PRO', *Linguistic Inquiry* 22, 647-686.

Koopman, H.: 1993, 'The Internal and External Distribution of Pronominal DPs', this volume.

Koopman, H.: 1994a, 'Licensing Heads', In N. Hornstein and D. Lightfoot, eds., *Verb Movement*, Cambridge University Press.

Mc Closkey, J.: 1992, 'On the Scope of Verb Movement in Irish', unpublished ms., University of California, Santa Cruz.

Morin, Y-C.: 1979, 'There is No Inversion of Subject Clitics in Modern French', ms., Université de Montréal.

Morin, Y-C.: 1985, 'On Two French Subjectless Verbs voici and voilà', *Language* 61.4.

Noonan, M.: 1989, 'Operator Licensing and the Case of French Interrogatives', In *Proceedings of the Eighth West Coast Conference on Formal Linguistics*, 8, 315-330.

Noonan, M.: 1992, 'Case and Syntactic Geometry', Doctoral Dissertation, McGill University, Montreal, Canada.

Perlmutter, D.: 1972, *Deep and Surface Structure Constraints in Syntax*, Holt, Rinehart and Winston, New York.

Pollock, J Y.: 1990, 'Verb Movement, Universal Grammar and the Structure of IP', *Linguistic Inquiry* 20, 365-424.

Renzi L. and L. Vanelli: 1982, 'I Pronomi Sogetto in Alcune Varieta Romanze', In Studi Linguistici in Onore di G.B,. Pelligrini, Pacini, Pisa.

Rizzi, L.: 1986, 'On the Status of Subject Clitics in Romance', In O. Jaeggli and C. Silva Corvalan, eds., *Studies in Romance Linguistics*, Foris Publications, Dordrecht.

Rizzi, L. and I. Roberts: 1989, 'Complex Inversion in French," *Probus* 1, 1-30.

Roberge, Y.: 1990, *The Syntactic Recoverability of Null Arguments*, McGill-Queen's University Press, Montreal.

Safir, K.: 1985, *Syntactic Chains*, Cambridge University Press.

Shlonsky, U.: 1988, 'Complementizer Cliticization in Hebrew and the Empty Category Principle', *Natural Language and Linguistic Theory* 6, 191-205.

Sportiche, D.: 1989, 'Le Mouvement Syntaxique: Constraintes et Paramètres', *Languages* 95, 35-80.

Sportiche, D.: 1983, *Structural Invarience and Symmetry in Syntax*, Doctoral Dissertation, MIT, distributed by the MIT Working Papers in Linguistics.

Sportiche, D.: 1988, 'A Theory of Floating Quantifiers and its Corollaries for Constituent Structure', *Linguistic Inquiry* 19, 425-449.

Sportiche, D.: 1995a, 'Clitic Constructions', In L. Zaring and J. Rooryck, eds., *Phrase Structure and the Lexicon*, Kluwer Academic Publishers, Dordrecht.

Sportiche, D.: 1995b, 'Sketch of a Reductionist Approach to Syntactic Variation and Dependencies', In H. Campos and P. Kempchinsky, eds., *Evolution and Revolution in Linguistics Theory*, Georgetown University Press.

Sportiche, D.: 1995c, 'French Predicate Clitics and Clause Structure', In A. Cardinaletti and M.T. Guasti, eds., *Small Clauses*, Syntax and Semantics, volume 28, Academic Press, New York.

Sportiche, D.: to appear, 'Pronominal Clitic Dependencies', In Henk van Riemsdijk, ed., *Language Typology: Clitics in the European Languages*, Mouton de Gruyter, Berlin.

Suñer, M.: 1991, 'Two Properties of Clitics in Clitic-Doubled Constructions', In J. Huang and R. May, eds., *Logical Structure and Linguistic: Cross-Linguistic Perspectives*, Reidel, Dordrecht.

Szabolcsi, A.: 1983, 'On the non-Unitary Nature of Verb-Second', ms., Max Planck Institute for Psycholinguistics, Nijmegen.

WORD ORDER IN SPANISH AND THE NATURE OF NOMINATIVE CASE*

Maria Luisa Zubizarreta
University of Southern California

Much of the research on word order in Romance languages lump Spanish and Italian together and contrast them with French. Spanish and Italian, but not French, are said to allow "free inversion" of the subject. This view is based on the fact that "subject post-posing" in French is only possible in the presence of a triggering element, namely an operator in Comp (a wh-word or its trace) or a subjunctive mood (which might also be analyzed as an operator in Comp). Such a construction, illustrated in (1c-e), is known as stylistic-inversion.[1] On the other hand, "subject post-posing" in Italian and Spanish does not need any triggering element, as shown in (2) and (3):

(1) a. *Marie* a téléphoné.
'M. phoned.'

b. *A téléphoné *Marie*.
'Phoned M.'

c. Je me demande *quand* a téléphoné *Marie*.
'I wonder when phoned M.'

d. *Quand* dis-tu qu'a téléphoné *Marie*?
'When did you say phoned M.'

e. Pierre voudrait que téléphone *Marie*.
'Pierre would like that telephone(subj.) M.'

(2) a. *Maria* ha telefonato.
'M. phoned.'

b. Ha telefonato *Maria*.
'Phoned M.'

(3) a. *Maria* ha llamado.
'M. called.'

b. Ha llamado *Maria*.
'Called M.'

K. Johnson and I. Roberts (eds.), Beyond Principles and Parameters, 223-250.

A closer look at the data reveals that Spanish differs from Italian in significant ways with respect to word order. As noted in Rizzi (1991), subject post-posing in Italian is sensitive to the nature of the object. While the sentences in (4) are marginal, the sentences in (5), where the direct object is cliticized onto the verb, and the sentences in (6), where the direct object is relativized, are perfect.[2]

(4) a. ?Ha risolto il problema *Gianni*.
'Has solved the problem Gianni'

b. ?Ha vinto la corsa *Gianni*.
'Has won the race Gianni'

(5) a. *Lo* ha risolto *Gianni*.
'It+has solved Gianni.'

b. *L*'ha vinta *Gianni*.
'It+has won Gianni.'

(6) a. Il problema che ha risolto *Gianni*
'The problem that has solved G.'

b. La corsa che ha vinto *Gianni*
'The race that has won G.'

The VOS order can also be rendered fully acceptable in Italian if the direct object is an indefinite (as noted in Belletti & Rizzi (1988))[3]:

(7) a. ?Ha scritto *la lettera Gianni*
'Has written the letter G'

b. Ha scritto *una lettera Gianni*
'Has written a letter G.'

The French Styl-Inv construction is also sensitive to the material inside the VP (cf. Kayne (1972)). More precisely, if an NP direct object is present, inversion is impossible, as shown in (8). But if the direct object is cliticized, the sentence improves significantly (cf. (9)); and if the direct object is relativized, the sentence becomes perfect (cf. (10)).[4]

(8) *Dites-nous quand voudrait lire *ces livres votre fils.*
'Tell us when would like to read those books your son.'

(9) a. ?Dites-nous quand *les* lira *votre fils*?
'Tell us when them+will read your son.'

b. Dites-nous quand voudrait *les* lire *votre fils*.
'Tell us when would like them+to read your son.'

(10) Montres-nous *les livres* que lira *votre fils*.
'Show us the books that will read your son.'

On the other hand, whether the direct object has been cliticized or relativized and whether the direct object is definite or indefinite, the VOS order is equally possible in Spanish. (11) are as acceptable as (12) and (13), and (14a) is acceptable as (14b).

(11) a. Ha resuelto el problema *Juan*.
b. Ganó la carrera *Juan*.

(12) a. *Lo* ha resuelto *Juan*.
b. *La* ganó *Juan*.

(13) a. El problema que resolvió *Juan*
b. La carrera que ganó *Juan*

(14) a. Escribió *la carta Juan*.
b. Escribió *una carta Juan*.

The data discussed above suggests that in Italian and in the French Styl-Inv construction, the post-verbal subject is within the VP. The post-posed subject is therefore sensitive to the nature of the material within the VP. On the other hand, the Spanish data is perfectly compatible with the hypothesis that the post-verbal subject is not within the projection of the verb, but rather within the projection of a functional category above VP.

There is evidence that suggests that the post-verbal subject is within the VP in Italian. Belletti & Rizzi (1981) note the following contrasts with respect to the distribution of bare numerals in Italian: a bare numeral in preverbal subject position gives rise to a well-formed sentence (see (15b) and (17b)), but a bare numeral in post-verbal subject position (see (16b)), like a bare numeral in object position (see (18b)), gives rise to an ill-formed sentence. This suggests that a post-verbal subject, like the object, is governed by the verb in Italian.[5]

(15) a. *Tre ragazze* hanno parlato.
'Have spoken three girls.'

b. *Tre* hanno parlato.
'Three have spoken.'

c. **Tre ne* hanno parlato.
'Three of them+have spoken.'

(16) a. Hanno parlato *tre ragazze*.
'Have spoken three girls.'

b. *Hanno parlato *tre*.
'Have spoken three.'

c. **Ne* hanno parlato *tre*.
'Of them+have spoken three.'

(17) a. *Tre settimane* sono passate.
'Three weeks have elapsed.'

b. *Tre* sono passate.
'Three have elapsed.'

c. **Tre ne* sono passate.
'Three of them+have elapsed.'

(18) a. Sono passate *tre settimane*.
'Have elapsed three weeks.'

b. *Sono passate *tre*.
'Have elapsed three.'

c. *Ne* sono passate *tre*.
'Of them+have elapsed three.'

Similar evidence in favor of the hypothesis that the subject is adjoined to the VP in the Stylistic Inversion construction is provided by Kayne (1981).

(19) a. Je me demande où *trois* sont partis.
'I wonder where three have left.'

b. *Je me demande où sont partis *trois*?
'I wonder where have left three.'

(20) a. Je me demande quand *trois* ont téléphoné?
'I wonder when three have phoned.'

b. *Je me demande quand ont téléphoné *trois*?
'I wonder when have phoned three.'

The well-known contrast between (21b) and (21c,d), with respect to the extraction of *combien*, also points towards the same conclusion. *Combien* can be extracted from a post-verbal subject or an object, but not from a preverbal subject.

(21) a. Tu crois que *combien de filles* ont téléphoné?
'You believe that how many of girls have phoned?'

b. **Combien* crois-tu que *de filles* ont téléphoné?
'How many do you believe that of girls have phoned?'

c. *Combien* crois-tu qu'ont téléphoné *de filles*?
'How many do you believe that have phoned of girls?'

d. *Combien* crois-tu que sont arrivés *de filles*?
'How many do you believe that have arrived of girls?'

Similar phenomena do not exist in Spanish and therefore we cannot compare Spanish with French and Italian in this respect. But there are a couple of contrasts between Italian and Spanish that suggest that the post-verbal subject is outside the VP in Spanish:

1. the distribution of temporal adverbs;
2. the possibility for a post-verbal subject to control the subject of a sentential adverb.

Rizzi (1991) notes that aspectual adverbs, but not temporal adverbs, can intervene between the verb and the post-posed subject in Italian. Compare (22) with (23). ((23) must be judged with a normal intonational contour.)

(22) a. Vince *sempre* Gianni
'Wins always Gianni.'

b. Non parla *più* nessuno
'Not speaks anymore anyone.'

(23) a. ??Ha telefonato *ieri* Gianni
'Has telephoned yesterday Gianni.'

b. ??Ti contatterà *domani* Gianni
'You will contact tomorrow Gianni.'

Similar contrasts are found in French:

(24) a. A qui parle *souvent* Jean?
'To whom speaks often J.?'

b. A qui ne parlera *plus* Jean?
'To whom not speaks anymore J.?'

(25) a. ??A qui a téléphoné *hier* Jean?
'Whom phoned yesterday J.?'

b. ??Qui contactera *demain* Jean?
'Whom will contact tomorrow J.?'

Let us assume that modifiers must be immediately dominated by a projection of the category that it modifies.[6] Aspectual adverbs modify the verb or the VP and must therefore be generated within the immediate scope of the verb (i.e. within the projection of V). On the other hand, temporal adverbs modify the tense morpheme located in TP and must therefore be generated within the immediate scope of Tense (i.e. within the projection of Tense). (Following current work, we assume that inflectional morphemes are attached to the verb via verb raising, at least in the Romance languages.) The contrast between (22) and (23) could therefore be taken as evidence that the post-verbal subject in Italian is dominated by VP (rather than by a higher functional projection). If the post-verbal subject in Italian and French is dominated by VP, it follows that temporal adverbs cannot precede the post-verbal subject in these languages.[7] Interestingly, in Spanish, the temporal adverb may precede the post-verbal subject. The Spanish counterparts of (23) are perfectly well-formed. Thus, there is no contrast between (26) and (27). This could be taken as evidence that the post-verbal subject in Spanish (unlike the post-verbal subject in Italian and French) is immediately dominated by a higher functional projection, rather than by VP.

(26) a. Gana *siempre Juan.*
b. No habla *más nadie.*

(27) a. LLamó *ayer Juan.*
b. Te contactará *manana Juan.*

Rizzi (1987) notes that a post-verbal subject in Italian cannot control the subject of a gerundive clausal adjunct.

(28) a. [PRO avendo terminato di mangiare], *Gianni* è andato via senza dire una parola.
'Having finished eating, G. left without saying a word.'

b. *[PRO avendo terminato di mangiare], è andato via *Gianni* senza dire una parola.

(29) a. [PRO avendo esaminato el caso, *il giudice* pronunciò un discorso.
'Having examined the case, the judge pronounced the sentence.'

b. *[PRO avendo esaminato il caso, pronunciò un discorso *il giudice*.

We may attribute the ungrammaticality of (28b)/(29b) to the fact that the post-verbal subject is too deeply embedded to function as the controller of the sentential adverb. More precisely, an NP within the VP may not control the subject of an adverb adjoined to S. This generalization is further supported by the ill-formedness of the following English and French sentences: **Having finished eating, it was suggested to John that he should leave/*Ayant fini de manger, il a été suggéré à Jean de partir*. In Spanish, the post-verbal subject may control the subject of a gerundive clausal adjunct. Once more, we may take this as an indication that the post-verbal subject in Spanish is not dominated by VP; it is dominated by a higher functional projection.

(30) a. [PRO habiendo terminado de comer], *Juan* se marchó sin decir una palabra.
'Having finished eating, J. left without saying a word.'

b. [PRO habiendo terminado de comer], se marchó *Juan* sin decir una palabra.

(31) a. [PRO habiendo examinado el caso], *el juez* dictó la sentencia.
'Having examined the case, the judge pronounced the sentence.'

b. [PRO habiendo examinado el caso], dictó la sentencia *el juez*.

On the basis of the above discussion I would like to suggest that Spanish and Italian (both *pro*-drop languages with "free subject inversion") differ with respect to the status of their post-verbal subject. In the VOS order the subject is immediately dominated by a functional category (higher than VP) in Spanish, while in Italian (as in the French Styl-Inv), the subject is immediately dominated by the VP.

Spanish differs from Italian in another important way. Spanish allows VSO order, both in matrix and embedded clauses.[8, 9]

(32) a. Escribió *Juan* la carta.
'Wrote J. the letter.'

b. Me trajo *Juan* el paquete.
'To me brought J. the parcel.'

c. Tiene *Juan* el libro.
'Has J. the book.'

(33) a. Es verdad que tiene *Juan* el libro
'It is true that has Maria the book.'

b. Me imagino que aceptó *Maria* la propuesta
'I imagine that accepted Maria the proposal.'

Italian does not allow VSO order in declarative, indicative sentences (whether matrix or embedded).

(34) a. *Scrisse *Giovanni* la lettera.
'Wrote G. the letter.'

b. *Mangió *Giovanni* la mela.
'Ate G. the apple.'

c. * Ha *Giovanni* il libro.
(Has G. the book)

(35) a. *E' vero che scrisse *Giovanni* la lettera.
'It is true that wrote G. the letter.'

b. *E' vero che ha mangiato *Giovanni* la mela.
'It is true that ate G. the apple.'

c. *E' vero che ha *Giovanni* el libro.
'It is true that has G. the book.'

Nor does French Styl-Inv allow the VSO order:

(36) a. *Quand a écrit *Jean* la lettre?
'When wrote J. the letter?'

b. *Quand dis-tu qu'a écrit *Jean* la lettre?
'When did you say that wrote J. the letter?'

Note that the post-posed subject may precede a PP complement in French Styl-Inv:

(37) On aime bien le cadeau qu'a envoyé Jean à Marie
'We like the present that sent Jean to Marie.'

Kayne (1986 Appendix II) provides evidence that in such cases the post-verbal subject is adjoined to the VP and the PP complement is extraposed. The evidence is provided by the contrast between (37) and (38):

(38) ??On connait l'endroit où a parlé Jean à Marie
'We know the place where spoke Jean to Marie.'

The sentences in (37) and (38) are derived via extraposition of the dative complement to the right, as shown in (39) and (40) respectively:

(39) le cadeau$_2$ que *pro* a envoyé $[e]_2$ $[e]_1$ *Jean* [à Marie]$_1$

(40) l'endroit$_2$ où *pro* a parlé $[e]_1$ $[e]_2$ *Jean* [à Marie]$_1$

In (40), but not in (39), there is a crossing of chains *1* and *2*. The marginality of (30) can therefore be attributed to the Crossing Constraint (first discussed by Bordelois (1974); see also Pesetsky (1982)).[10] The Crossing Constraint can also explain the contrast between (41) and (42), assuming the basic order to be *plaindre de NP à NP*:

(41) Pierre connaissait la fille de qui s'est plaint Jean à Marie.
'P. knew the the girl of whom complained Jean to Marie.'

(42) ??Pierre connaissait la fille à qui s'est plaint Jean de Marie.
'P. knew the girl to whom complained Jean of Marie.'

Interestingly, the same contrasts are found in Italian:

(43) a. A Gianni piace il regalo che ha mandato Francesco a Maria.
b. ??Si conosce il luogo in cui ha parlato Gianni a Maria.

(44) a. Gianni conosceva la ragazza della quale si è lamentato Francesco con Maria.
b. ??Gianni conosceva la ragazza con la quale si è lamentato Francesco di Maria.

But, as shown below, comparable contrasts are nonexistent in Spanish, which suggests that such structures are not derived via PP extraposition in this language:

(45) a. Nos gusta el regalo que envió Juan a Maria.
b. Conocemos el lugar en que habló Juan a Maria.

(46) a. Conocemos a la persona de quien se quejó Juan a María.
b. Conocemos a la persona a quién se quejó Juan de María.

The same kind of contrasts are found in interrogatives in French and Italian, but not in Spanish. Compare (47) and (48) with (49).

(47) a. Je me demande quel cadeau a envoyé Jean à Marie.
b. ??Je me demande où a parlé Jean à Marie.

(48) a. Mi domando che regalo ha mandato Gianni a Maria.
b. ??Mi domando dove ha parlato Gianni a Maria.

(49) a. Me pregunto cuál regalo envió Juan a María.
'I wonder which present sent J. to M.'

b. Me pregunto donde habló Juan a María.
'I wonder where talked J. to M.'

On the basis of the above discussion, we conclude that the V-S-PP order, as well as the VSO order, are base generated structures in Spanish. The issue of whether the subject in these structures is immediately dominated by VP or by a higher functional category is a more complex one. Since Italian lacks the VSO order altogether, it is extremely difficult to find relevant contrasts between the two languages that could shed light on this question (as we did in examining the VOS order). But if the control test discussed earlier is of any significance, the grammaticality of sentences like (50) suggests that the subject in the VSO order is also dominated by a functional category higher than VP:

(50) Habiendo terminado de examinar el caso, dictó *el juez* la sentencia.
'Having finished examining the case, pronounced the judge the sentence.'

My tentative conclusion is then that the subject is immediately dominated by a functional category higher than VP in the VSO order, as well as in the VOS order.

1.

In this section I will provide an analysis that can account both for the VSO and VOS order in Spanish. But I will first attempt to answer the following, fundamental question: to what other grammatical property of the language could we link the existence of the VSO order? My proposal is that the existence of VSO in Spanish and the lack of it in Italian should be linked to another salient difference between the two languages: in Spanish, but not in Italian, an overt morphological Case marker (i.e. the Case marker *a*) distinguishes the object from the subject in a number of crucial cases. We explain below.

In Spanish we find the paradigm in (51) (irrespective of word-order). Specific, animate objects (whether definite, indefinite, or quantified) are preceded by the Accusative Case marker *a*, as illustrated in (51a-c). The animate nonspecific negative polarity item *nadie* is also preceded by the Acc Case marker *a*, as illustrated in (51d).

(51) a. Juan vió *a* Maria.
'J. saw *a* M.'

b. Juan busca *a* una muchacha (que sabe hablar inglés).[11]
'J. seeks *a* girl (that can speak (ind) English)'

c. Juan observó *a* cada candidato.
'J. observed *a* each candidate.'

d. Juan no vió *a* nadie.
'J. didn't see *a* anybody.'

Interestingly enough, the Case marker *a* also appears with inanimate objects in some cases (as noted in Lois (1989). These are cases in which the subject is also inanimate. In some cases the presence of the ACC Case marker *a* is optional, as illustrated in (52). But in other cases, such as those in (53), the ACC case maker is obligatory.

(52) a. Los ácidos atacan (*a*) los metales.
'The acid attacks the metal.'

b. El entusiasmo vence (*a*) la dificultad.
'Enthusiasm defeats difficulties.'

c. Un crudo invierno azota (*a*) esta región.
'A harsh winter lashes this region.

d. La tormenta hundió (*a*)l barco.
'The storm sank the ship.'

(53) a. El sujeto precede *a*l objeto.[12]
'The subject precedes/follows *a* the object.'

b. El otono sigue *a*l verano.
'The fall follows the summer.'

More precisely, in the case of noncausative/nonagentive verbs with an inanimate subject and an inanimate object, the object must be preceded by the Acc Case marker *a*. On the basis of this observation, we may hypothesize that Nominative Case (although morphologically absent) is distinctive in Spanish. Nominative Case is distinctive to the extent that it is morphologically distinct from the Accusative Case or to the extent that it is distinguished from the Accusative with respect to prominent selectional features: the Nominative is the "causer" and the Accusative is the "patient" or the Nominative is "animate" and the Accusative is "inanimate". This property of the Nominative Case (of being distinctive) constitutes a crucial difference between Spanish and Italian, and it is this property which licenses the VSO order in Spanish.

The next question is how can we relate formally the above-mentioned generalization about Spanish (NOM Case is distinctive) with the distribution of subjects. In his study of the history of the clausal structure in French, Roberts (1991) shows that there is an interesting link between word-order in embedded clauses and morphological Nom Case in Old French (12th and 13th century). And he proposes to capture this correlation by assuming that in certain languages NOM Case can give rise to an independent functional projection. It is this basic insight that we would like to adopt here.

A few preliminary assumptions need to be spelled out. First, we adopt Pollock's (1989) assumption that tense is a functional category independent of Agr (subject agreement), and that TP is located between CP and AgrP. This latter assumption seems natural in light of the fact that the choice of complementizer and of tense is intimately related; there is an interdependency between Comp and tense of a selectional nature. For example, the complementizer *that* selects [+tense]; the complementizer *for* selects [-tense]. Second, it is clear that the presence of NOM Case is intimately related to the presence of Tense. This can best be appreciated in Spanish, where infinitival complements do not allow for a lexical subject, but where infinitival clausal adjuncts do allow for a lexical subject. Cf. *Al llamar esta manana tu madre, me dí cuenta que algo grave había pasado*. 'When call(inf) this morning your mother, I realized that something serious had happened'. Rigau (1992) and Elordieta (1992) argue that such infinitival clausal adjuncts are temporally independent: they have their own tense specification. Hence, we can assume that tense, although morphologically absent, is semantically and syntactically present in such infinitival adjuncts. In many languages, the T(ense) category that carry the NOM Case feature is borne by the same category that bears the tense feature. But, elaborating on Roberts' proposal, we would like to make the following suggestion:

(54) If NOM Case is distinctive (in the sense defined above) in a given language, it may be borne by a functional projection distinct from the functional category that carries the tense feature.

In discussing the nature of grammatical positions, Mahajan (1990) and Chomsky & Lasnik (1991) distinguish between L-related and non-L-related positions (a distinction which subsumes the earlier A/A-bar distinction). L-related positions are those that are contained within the projection of a lexical category (i.e. positions that are thematically related to a head) or within the projection of a functional category morphologically related to a lexical category. Thus, the Spec of Agr and the Spec of Tense are defined as L-related, to the extent that Agr and tense are morphologically attached to the verb. Chomsky (1992) (within a somewhat different conception of morphological and Case relations) introduces a further refinement in the notion of L-related, distinguishing between the notions of N-related and V-related. In particular, he notes that Tense is both V-related, to the extent that it carries the tense feature attached to the verb, and N-related, to the extent that it carries the NOM Case feature assigned to (or copied onto) the subject NP. We can view our proposal in (54) as a case in which the N-related function of T is syntactically split from the V-related function of T. Or to put it in a slightly different way, in languages like Spanish, T_V (for V-related) and T_N (for

N-related) constitute a discontinuous category. This gives us the representation in (55) for the clause structure in Spanish, in which T_V and T_N are split. (We assume that T_V is immediately dominated by $\overline{C}$ because, as we remarked earlier, Comp and tense specifications are dependent. Irrelevant details, such as AgrO, are omitted.)

(55)

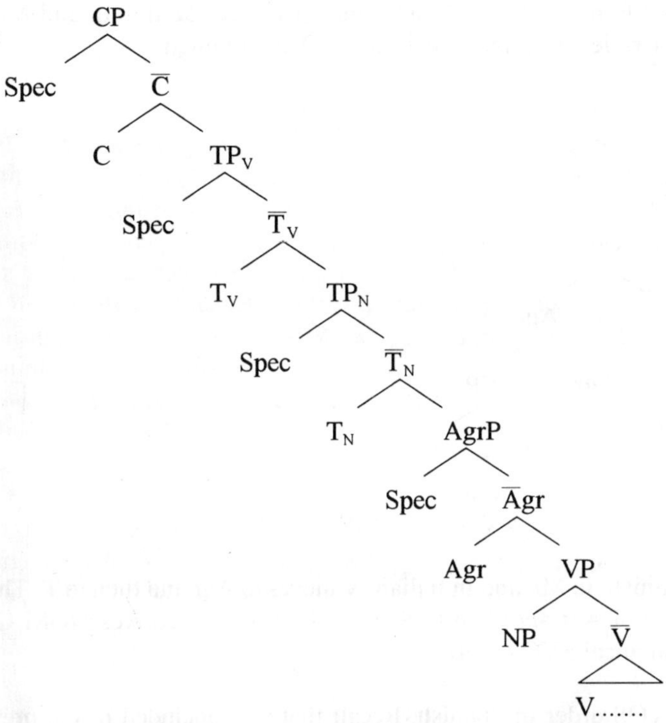

V moves to Agr, then to T_N, and finally to T_V (thus picking up the Agr and tense features). The NP subject moves to Spec of Agr (undergoing agreement) and then to Spec of T_N. We assume (with Mahajan (1990), Koopman & Sportiche (1991), Roberts (1991) and others) that there are two syntactic configurations for Case assignment: via government and via the Spec-head relation. Recall that T_N carries nominative Case; therefore its Spec is a Case position. It is therefore in Spec of T_N that the NP subject receives Nominative Case. This gives rise to the VSO order in Spanish.[13]

The SVO order is obtained if the NP subject raises from Spec of T_N to Spec of T_V. Note that NP movement to Spec of T_N is motivated by Case: it is movement from a non-Case position to a Case position. On the other hand, NP movement to Spec of T_V in the SVO order is not motivated by Case: it is movement from a Case position to a non-Case position. If anything, this movement (which is optional) is triggered by semantic considerations: Spec of T_V is a topic position and preverbal subjects are the unmarked topic in Spanish (see note 3). If Spec of T_V is not a Case position, we might expect that phrases other than the subject may move into this position. As we will see in the next section, this prediction is borne out. Movement to Spec of T_V is topicalization in Spanish.

In Italian and Modern French (MF) NOM is not distinctive. Therefore, the V-related and the N-related functions of tense cannot be split. There is only one T node (which is both V-related and N-related). The basic clause structure for Italian and Modern French is as in (56) (irrelevant details, such as ArgO, are omitted).

(56)

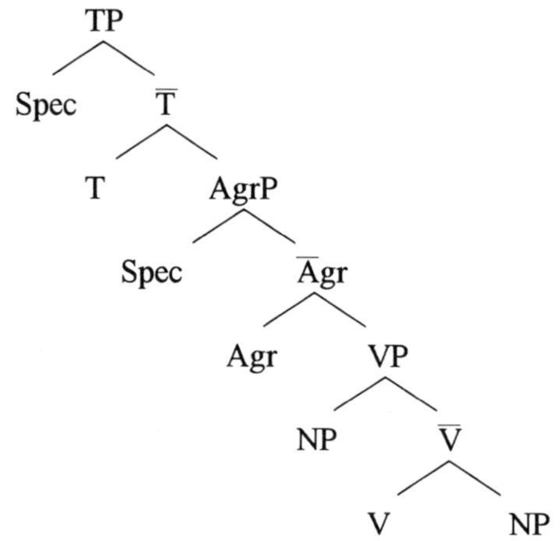

As in the case of Spanish, in MF and in Italian, V moves to Agr and then to T. The NP subject raises to Spec of Agr and then to Spec of T (where it receives NOM Case), giving rise to the canonical SVO order.

Consider next the VOS order in Spanish. Recall that we concluded in the previous section (based on data having to do with the distribution of temporal adverbs and with subject control) that the post-verbal subject is dominated by a functional category higher than VP. What can this position be? There are three Spec positions, which could in principle be generated to the right of VP: Spec of Agr, Spec of T_N, and Spec of T_V. Spec of Agr is excluded because it is not a Case position. It cannot be Spec of T_V either because this is a topic position and post-verbal subjects can never be interpreted as topics (see footnote 3). Therefore, the subject in the VOS order in Spanish must be in Spec of T_N. The sentences in (11)-(14) have the structure in (55), but with Spec of T_N generated on the right of T_N.

Now why may the Spec of T_N, but not the Spec of T_V, be generated to the right of its head? Let us assume that the canonical position of the Spec of a V-related category is adjacent to its head:

(57) a. b.

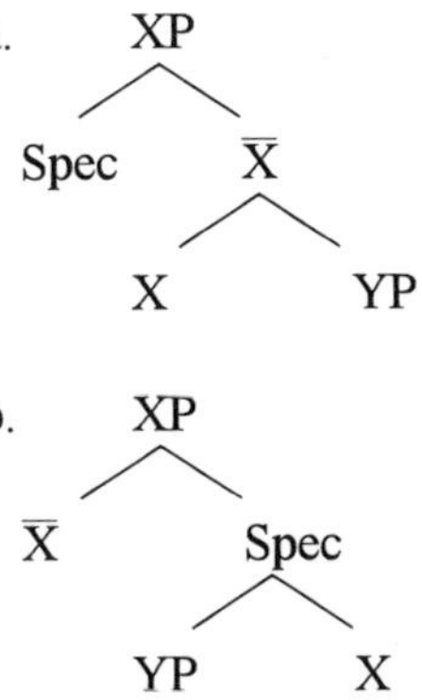

(Where X = V-related category)

In other words, the Spec of a V-related head is identifiable in the structure (or more precisely in the PF structure if the linear order is relevant to PF rather than to S-Structure or LF) to the extent that it is adjacent to its head. On the other hand, there is no adjacency constraint on the Spec of an N-related head; the Spec of an N-related head does not rely on adjacency to be identified.[14] We may speculate that this is so because such a Specifier bears a feature of its head. In the case under consideration, Spec of T_N bears the NOM feature of T_N (more precisely, the Case of T_N is copied on the N in its Spec). Therefore, the Spec of an N-related category is identified by virtue of sharing the same feature as its head, rather than by virtue of being adjacent to its head.

In section 1 we argued that the post-verbal subject in Italian and MF is inside the VP, rather than within a functional projection higher than the VP. Recall that Italian and MF lack the split T. In these languages, T is therefore a V-related category, as well as an N-related category. Given the proposal in the previous paragraph, it follows that these languages cannot tolerate the Spec of T to be generated to the right of T. Note that Agr is also both a V-related and a N-related category (it is related to V to the extent that Agr is morphologically borne by V and it is N-related to the extent that it enters into an agreement relation with N). If the Spec of a V-related category must be adjacent to its head, then Spec of Agr cannot be generated to the right of Agr in a head initial language like the Romance languages. The post-verbal subject in Italian and MF must therefore be analyzed as occupying its D-Structure position, namely the Spec of V. It is necessary to assume furthermore that NOM Case can be assigned to the Spec of V in these languages via the trace of the verb (which has raised to T). If the proposal we have made above is correct (i.e. the Spec of V-related categories must be adjacent to its head), then it must be the case that in transitive structures such as (4)-(7) the subject must be in pre-head position. In other words, the Spec of V cannot be generated to the right of V. The only way that the VOS order can be generated in Italian then is by raising V to T and raising the object to Spec of Agr_O (Agr_S stands for subject agreement and Agr_O stands for object agreement).

(58)

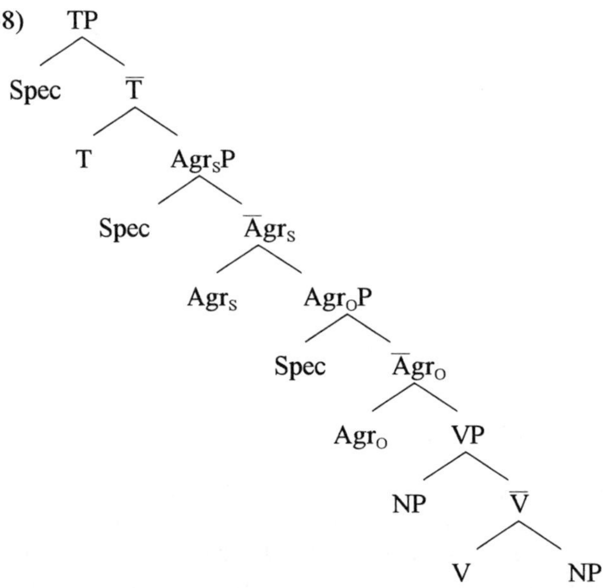

This analysis can perhaps gives us the basis to describe the constraints on the coexistence of a post-verbal subject and a direct object in Italian; namely, in terms of which kinds of NPs can move into Spec of Agr_O across a VP internal subject. We leave this question open for further research.[15]

2.

Given the analysis outlined in section 2, one expects that phrases other than the subject may occupy the Spec of T_V position in Spanish. Spec of T_V is not a Case position. Thus, if the subject does not raise, we would expect that a constituent other than the subject could be moved to that position. Interestingly, the prediction is borne out. Piera (1987) provides the following data, in which various kinds of adverbial elements are found in preverbal position (with no pause or intonation change between the adverbial and the verb):

(59) a. *Temprano* salía *Julía* de casa.
'Early left J. the house.'

b. *Por la noche* salía *Julía* de casa.
'During the night went out J. from the house.'

c. *En esta ciudad* salía *Julía* todas las noches.
'In this city went out J. every night.'

d. *Con poco dinero* salía *Julía* de casa.
'With little money went out J. from the house.'

e. *Cansada* salía *Julía* de casa.
'Tired went out J. from the house.'

(60) a. Supe que *bajo ningún pretexto* conseguiría *Juan* acercarse a nosotros.
'I knew that under no conditions would+manage J. to approach us.'

b. *Limpia como una patena* tenía *Eduardo* la casa
'As clean as a whistle had E. the house.'

c. *Cantando en la ducha* estaba *Edelmira* cuando llegué.
'Singing in the shower was E. when I arrived.'

Piera notes furthermore that a preposed adverbial is in complementary distribution with a preposed subject:

(61) a. *Temprano (con poco dinero/cansada...) Julía salía de casa.
b. *Limpia como una patena Eduardo tenía la casa.
c. *Cantando en la ducha Edelmira estaba cuando llegué.

(61a) contrasts with (62), in which there is an intonation break between the adverb and the subject. In (62) the adverb is adjoined to the clause.

(62) Temprano (con poco dinero/cansada...), Julía salía de casa.

Piera also notes that adverbial preposing is not permitted in infinitival clauses. (Again, care should be taken not to introduce an intonational break after the adverb.)

(63) a. Salir de casa *en esta ciudad* es malo para la salud.
b. **En esta ciudad* salir de casa es malo para la salud.

(64) a. Salir de casa *con poco dinero* es malo para la salud.
b. **Con poco dinero* salir de casa es malo para la salud.

(65) a. Salir de casa *cansado* es malo para la salud.
b. **Cansado* salir de casa es malo para la salud.

The ungrammaticality of such forms follows directly from the analysis proposed here. Infinitival clauses lack a split T node because a [-tense] category is incompatible with NOM Case. As in French and Italian, in Spanish the Spec of T in such infinitival clauses is presumably occupied by subject PRO. In other words, to the extent that a preverbal topic position is associated with a split T in Spanish, and that a split T is contingent on the presence of NOM Case, then it follows that if a lexical NP subject is unavailable, then fronting of constituents should also be unavailable. Thus, we can capture the generalization that infinitival clauses do not allow adverb preposing for the same reason that it does not allow for overt lexical subjects.[16] [17]

Let us now turn to preposed arguments. PP complements may also be readily preposed.

(66) a. *Con María* hablaron los padres.
'With M. talked the parents.'

b. *De María* se quejó la maestra.
'Of M. complained the teacher.'

On the other hand, the preposing of objects is more complex. Meyer (1972) notes that OSV and OVS are not possible orders. The same point is made by Contreras (1991). ((67b) must be judged with a flat intonation.[18])

(67) a. **La carta Juan* escribió.
b. **La carta* escribió *Juan*.

The ungrammaticality of (67a) is readily explained because the preposed object and the preposed subject are competing for the same position (i.e. Spec of T_V).

The ungrammaticality of (67b) is more puzzling, especially in light of the grammaticality of (68b). In effect, if the object is clitic doubled, the OVS order becomes possible. (Again, (68a) should be judged with no intonational break or change between the fronted object and the subject.)

(68) a. **La carta$_1$ Juan la$_1$* escribió.
b. *La carta$_1$ la$_1$* escribió *Juan*.

Our claim is that (68b) is a case of NP movement to Spec of T_V (i.e. a case of local fronting to an A-position), and it is not to be confused with cases of left-dislocation. Hernanz & Brucart (1987) have shown that left-dislocation in Spanish (which they call "rematización") are base-generated structures (i.e. they do not obey Subjacency). The dislocated element is outside the CP and can be related to a gap across one or more embedded clauses. This explains the contrast between (69a) and (69b), where the latter has either an intonational break or an intonational change between the preposed object and the verb.

(69) a. *La carta$_1$ piensa Maria que la$_1$ escribió Pedro.
'The letter thinks M. that it+wrote P.'

b. La carta$_1$, piensa Maria que la$_1$ escribió Pedro.

As expected, a preposed object is not possible in an infinitival complement either:

(70) *Maria piensa la carta escribirla.
'M. thinks the letter write(inf)+it.'

A preposed indirect object patterns like the direct object. The IO-S-V order is impossible, and the IO-V-S is possible only to the extent that the IO is clitic-doubled.

(71) a. **A Juan Pedro* habló/escribió.
'To J. Pedro spoke/wrote.'

b. **A Juan* habló/escribió *Pedro*.
'To J. spoke/wrote Pedro.'

c. *A Juan$_i$ le$_i$* habló/escribió *Pedro*.
'To J. to him+habló/escribió Pedro.'

IO preposing is to be distinguished from left-dislocation. Unlike dislocation, IO preposing is a local process:

(72) a. **A Juan*$_1$ dice Maria que *le*$_1$ habló Pedro.
'To J. says M. that to him+spoke P.'

b. *A Juan*$_1$, dice Maria que *le*$_1$ habló Pedro.

IO preposing is also excluded from infinitivals:

(73) *Maria piensa *a Pedro* hablar*le*.
'M. thinks to P. speak(inf)+to him.'

Until now we have been relying on intonation to distinguish cases of preposing to Spec of T_V from cases of dislocation, a delicate test to say the least. Fortunately, in the case of complements, there is another test to distinguish between the two based on the distribution of negative polarity items. In Spanish, a negative polarity item in post-verbal position must be licensed by a negative element. Thus, a post-verbal subject patterns in this respect with a direct object, as shown in (74b) and (74c), and contrasts with a preverbal subject (see (74a)).

(74) a. *Nadié* (**no*) llamó.[19]
'Anybody not called.'

b. *(*No*) llamó *nadie*.
'Not called anybody.'

c. Juan *(*no*) llamó *a nadie*.
'Juan not called anybody.'

Interestingly, a preposed complement behaves in this respect just like a preposed subject. (75) illustrates the distribution of a negative polarity within a PP and (76) illustrates the distribution of a negative polarity IO.

(75) a. La maestra *no* habló *con nadie*.
'The teacher not talked with no one.'

b. *Con nadie* habló la maestra.
'With noone talked the teacher.'

(76) a. La maestra *no* le habló *a nadie*.
'The teacher not to him+talked to no one.'

b. *A nadie* le habló la maestra.
'To noone to him+talked the teacher.'

An accusative clitic may double a specific, animate direct object, but not a nonspecific NP like *nadie*:

(77) a. *A Pedro lo* vió/examinó/visitó María.
'P. ACC cl. saw/examined/visited M.'

b. ??*A nadie lo* vió/examinó/visitó María.
'Nobody ACC cl. saw M.'

But there is a dialect (called "leista") in which the direct object may be clitic-doubled with the dative clitic and, interestingly, in this dialect the dative clitic may double the direct object *a nadie*.[20]

(78) *A nadie$_1$ le$_1$* vió/examinó/visitó María.
'No one DAT cl. examined the doctor.'

The following examples show that *nadie* cannot be left-dislocated; they are unacceptable independently of whether or not there is an intonational break or change between *nadie* and the adjacent verb.[21]

(79) a. **Con nadie* piensa Pedro que habló María.
b. **A nadie* piensa Pedro que habló María.
c. **A nadie* piensa Pedro que vió María.

The contrast between (79), on the one hand, and (74a), (75b), (76b), (78), on the other hand, therefore shows that the preposed complements in the latter set of examples should not be analyzed as dislocation (independently of considerations of intonation), but rather as a local fronting phenomenon.

The question that remains to be answered is why direct and indirect objects cannot be fronted unless they are clitic-doubled. At first sight, this would appear to be a Specified Subject effect (subsumed under some version of the Relativized Minimality Condition; cf. Rizzi (1990), Aoun & Li (1993)): an NP object cannot move across the NP subject of its clause because the NP subject would function as the minimal potential antecedent. The clitic voids a Minimality violation. The exact implementation of this hypothesis will depend in part on how we choose to analyze clitics in Spanish.

Supposing Minimality is a condition on chains, we could interpret the role of the clitic as breaking down the chain of the fronted object into two subchains: (NP (cl, e)), such that the minimal antecedent for the object trace is now the clitic rather than the fronted object. This analysis relies on the view that clitics in Spanish are part of an argument chain.

A slightly different alternative which does not directly appeal to the presence of an intervening subject, nor on the view that clitics are arguments, could be the following. Suppose that dative and accusative clitics in Spanish are agreement morphemes (Suñer (1988), Franco (1990). More precisely, suppose that Agr_O is phonologically realized as a clitic when it is activated (that is when it undergoes agreement with a phrase in its Specifier). We can then interpret the obligatory presence of the clitic as indicating that the fronted object must move through the Spec of Agr_O (of dative Agr_O in the case of indirect objects and accusative Agr_O in the case of direct objects). This can attributed to a version of Minimality formulated as follows by Chomsky & Lasnik (1991):

(80) Minimize chain links.

This analysis is then based on the view that a clitic-doubled object in Spanish receives Case via Spec-head agreement, while a non clitic-doubled NP object receives Case via government. In the first case the object moves to Spec of Agr; in the second case the object is in its canonical position (sister to V). In short, a fronted object must move through Spec of Agr_O (triggering the spell out of Agr_O as a clitic) in order for the resulting chain to conform to (80).

3. CONCLUSION

In this paper we have suggested that the status of preverbal and post-verbal subjects in Spanish is different from that of Italian and French. In Spanish a subject can appear between the verb and the object, but not in Italian and French. In Italian and French, the post-verbal subject is sensitive to the material inside the VP, but not in Spanish. We provided an analysis which relates these properties of the Spanish subject to the nature of NOM Case in this language. In Spanish, NOM Case is distinctive and may therefore be borne by a distinct functional category. More precisely, the V-related and the N-related functions of the tense node are split in Spanish: T_V (which bears the temporal specification) and T_N (which bears NOM Case). The post-verbal subject in Spanish was argued to be located in the Spec of T_N, while the post-verbal subject in Italian and Modern French (which lack a split T) was argued to be inside the VP. Correlated with the existence of a split T in Spanish is the existence of a non-Case position at the beginning of the clause (the Spec of T_V), where subject, adverbs, and complements can move to. We suggested that Spec of T_V is the topic position in Spanish.

Given the definition of L-related category, Spec of T_V is an L-related category, but empirical arguments that could support this point are difficult to come by. One popular test to distinguish fronted phrases in an L-related position from fronted material in a non-L-related position is weak-cross over (see Mahajan (1990)): movement to an L-related position across a bound pronoun is permissible, but movement to a non-L-related position across a bound pronoun is not allowed. At first blush, Suñer's (1988) facts could be interpreted as supporting the view that Spec of T_V is an L-related position:

(81) a. *A cuales de los ninos*$_1$ no *los* aguanta ni *su*$_1$ madre?
'Which of the children not Acc cl+stand not even their mother?'

b. **A cuales de los ninos*$_1$ piensa *su*$_1$ madre que no *los* aguanta ni la profesora?
'Which of the children thinks their mother that not Acc cl+stand not even the teacher?'

If in (81a) the wh-phrase moves to Spec of T_V before moving to Spec of Comp and if Spec of T_V is an L-related position, then the lack of Weak Crossover effects in this sentence follows (i.e. the extraction of the wh-phrase across the bound pronoun is to an L-related position). On the other hand, in (81b) movement of the wh-phrase across the bound pronoun is directly to a non-L related position (i.e. the higher Comp), thus giving rise to a Weak Crossover violation. But, unfortunately, this analysis makes a prediction which does not seem correct. If the movement of the wh-phrase is through Spec of T_V in (81a), then we would expect to have obligatorily a post-posed subject in sentences of the type illustrated below. While many speakers prefer the (a) option (with a post-posed subject) for the bound reading, it is generally agreed that the bound reading of the pronoun in the (b) option (with a preverbal subject) is not excluded:

(82) a. *A cuales de los ninos*$_1$ dice la profesora que no *los* aguanta ni *su*$_1$ madre?
'Which of the children says the teacher that not Acc cl+stands not even their mother?'
b. *A caules de los ninos*$_1$ dice la profesora que ni *su*$_1$ madre no lo aguanta?
'Which of the children says the teacher that not even their mother not Acc cl+stands?'

Therefore the contrast in (81) cannot be taken as empirical evidence that Spec of T_V is an L-related position.

It is to be noted that although Spanish has the XP^VS order, it is clear that Spanish is not a V2 language (i.e it is not a language with a generalized V-to-Comp movement). In V2 languages there are root/non-root asymmetries with respect to word order. This is not the case in Spanish. Furthermore, in V2 languages the verb cannot appear in first position; some material must obligatorily move to Spec of Comp. Again, this is obviously not the case in Spanish. Roberts (1990) has argued that it is not possible to adjoin adverbs to sentences in which V-to-Comp movement has applied. In Spanish, it is possible to adjoin adverbs to clauses with the order XP^VS:

(83) a. Todas las noches, con Juana sale Pedro de juerga.
'Every night, with J. goes out P. for fun.'

b. Creo que ayer, de Juana se quejaron todos los profesores
'I believe that yesterday, of J. complained all the professors.'

That Spanish is not a V2 language does not mean, of course, that there are no cases of V-to-Comp movement. It has been argued that there is movement of V to Comp in the case of interrogatives (Rizzi (1990)). As is well known, a fronted wh-phrase requires verb preposing:

(84) a. (Nos preguntamos) *Qué* compró *María*?
'(We wonder) What bought Maria?'

b. (Nos preguntamos) **Qué María* compró?

It is unlikely that the analysis of XP^VS order in affirmative contexts can be extended to interrogative sentences such as the ones above (as suggested by Goodall (1991)). The reason is that Italian exhibits the same "inversion" phenomenon as Spanish in interrogatives, although the contrast is slightly weaker in embedded clauses than in matrix clauses in this language (see Rizzi (1991)):[22]

(85) a. **Che cosa il direttore* ha detto?
'What has the director said?'

b. *Che cosa* ha detto *il direttore*?
'What has said the director?'

(86) a. ??Tutti si domandano *che cosa il direttore* ha detto.
'Everybody wonders what the director has said.'

b. Tutti si domandano *che cosa* ha detto *il directore*.
'Everybody wonders what has said the director.'

On the other hand, as we have seen earlier, "inversion" in affirmative sentences is much more restricted in Italian than in Spanish. The analysis of "inversion" in interrogatives proposed by Rizzi (and which we adopt here) is that the Spec of an interrogative clause must agree with a [+QU] Comp. In many languages, among them Spanish and Italian, the QU feature is borne by the verb. The verb must therefore move to Comp in order to enter into an agreement relation with the Spec of Comp, which contains a wh-phrase or a yes/no operator.

Nevertheless, the "inversion" paradigm in (84) is interesting because it provides further support for the hypothesis that *XP* in *XP^VS* structures may be analyzed as occupying the same position as the preverbal subject in *SVO* structures. In effect, the contrast illustrated in (84) is found with an *XP* other than the subject:

(87) a. *(Me pregunto) *Qué a María* le envió Juan?
'(I wonder) what to M. DAT cl. sent J.?'

b. (Me pregunto) *Qué* le envió Juan a María?
'(I wonder) what DAT cl. sent J. to M.?'

(88) a. *(Me pregunto) *Qué ayer* compró Juan?
'(I wonder) what yesterday bought J.?'

b. (Me pregunto) *Qué* compró Juan ayer?
'(I wonder) what bought J. yesterday?'

It would be interesting to investigate whether the parametric difference between Spanish and Italian/French with respect to NOM Case assignment could be extended to ACC Case assignment. In other words, is ACC Case distinctive in Spanish? And if so, does ACC define an independent functional projection from Agr_O in Spanish? If yes, one piece of contrastive data that could be linked to this proposal is the one illustrated in (89). As is well known, in French and Italian, but not in Spanish, a clitic direct object may trigger past participle agreement.

(89) a. Pierre *l*'a écrit*e* (la lettre)
'P. it(fem.sing)+has written(fem.sing) (the letter).'

b. Piero *l*'ha scritt*a* (la lettera)

c. Pedro *la* ha escrito (la carta)

Another difference between Spanish and French/Italian that might be related to (89) is the existence of clitic-doubling in the former but not in the latter languages. We leave these questions open for future research.

NOTES

* I would like to thank V. Demonte, H. Contreras, and K. Zagona for written comments on an earlier version of this paper, as well as J. Aoun, E. Benmamoun, J.-R. Vergnaud, and the students in my Romance syntax class (Spring 1992) for valuable discussions. This paper was presented at the II Coloquium on Generative Grammar held at the Universidad del País Vasco, Vitoria-Gasteiz, April 1992.

[1] The French Sty-Inv is optional and has a stylistic flavor to it.

[2] Mario Squartini informs me that the VOS order in Italian improves considerably in the progressive aspect: *Sta scrivendo la lettera Gianni* (Is writing the letter Gianni.). More research needs to be done in this area.

[3] In fact, an indefinite postposed subject can also render the VOS order natural: *Ha scritto la lettera una studentessa* (Has written the letter a student.).

[4] But note that indefiniteness plays no role in licensing VOS in French Styl-Inv. Compare (i) with (ii).

(i) *Je me demande quand a écrit *la lettre Jean.*
(ii) *Je me demande quand a écrit *une lettre Jean.*

[5] Note the difference between a subject (whether preverbal or post-verbal) and an object with respect to *ne* cliticization: (15c), (17c), (16c) vs. (18c). The grammaticality of (18c) constitutes an argument for the unacccusative nature of *passare*.

[6] For discussion of the structural relation between modifiers and modifees, see Zubizarreta (1982).

[7] Note that for this argument to go through we need to make the further assumption that temporal adverbs are generated at the rightmost edge of TP. This assumption is sustained by the ungrammaticality of the following sentence:

(i) **Jean demain viendra.*
(John tomorrow will come.)

After V-to-T movement applies the temporal adverb must end up to the right of the verb.

[8] This is not to say that the three examples in (32) are equivalent. Preverbal and post-verbal subjects are associated with a different information structure. As shown in Meyer (1972), the preverbal subject in Spanish is the unmarked topic of the sentence. The post-verbal subject never functions as the topic, but rather as the focus.

Meyer (1972), reporting the judgements of a native speaker from Madrid, says that the VSO order in matrix clauses require an indefinite subject, while no such restriction on a VSO order exists in a subordinate clause:

(i) Se ha comido *un nino/*Paco* la manzana.
(Has eaten a boy/Paco the apple.)

(ii) Es verdad que se ha comido *un nino/Paco* la manzana.
(It is true that has eaten a boy/Paco the apple.)

No such contrast exists in the dialect that I am reporting here. My guess is that the contrast between the indefinite and the definite found in (i) for the dialect (or idiolect) reported by Meyer is to be attributed to differences in the fine-grained details of the information structure associated with the different word orders, a point which I cannot develop here.

[9] Pierce (1989) cites evidence from first language acquisition of Spanish speaking children that suggests that, at an early stage (18-24 months), there is little variability of word order, the SV order being preferred. But the 25-30 month period is rich in such variability, post-verbal subjects being very common especially with monadic verbs. The data seems to indicate furthermore that for transitive verbs the VSO order is more common than VOS at this stage.

[10] Presumably, only A'-chains (i.e. chains headed by a non-A(rgument) position) are visible to the Crossing Constraint.

Note that the chain between the preverbal subject position and the post-posed subject does not count for the Crossing Constraint. This suggests that the preverbal subject position has the status of a pronominal expletive and that the post-posed subject is an A(rgument) position. See Roberts (1991) for a defense of this view. But see Pollock (1986) for arguments to the contrary.

Note furthermore the example in (i) (from Kayne (1986)), or the interrogative counterpart in (ii):

(i) ??On connait le jour où a écrit Jean à Marie.
'We know the day when wrote J. to M.'

(ii) ??Je me demande quand a écrit Jean à Marie.
'I wonder when wrote J. to M.'

While it is probably the case that locatives are verbal complements, temporals are not. In fact, we argued earlier that they are modifiers of TP, and must therefore be analyzed as immediately dominated by Tense. This means that for the explanation of the marginality of (i) and (ii) in terms of the Crossing Constraint to go through, it must be the case that the extraposed PP is adjoined to TP, as shown in (iii) (ignoring intermediate functional projections such as AGR).

(iii) ...où$_1$ [$_{TP}$ *pro*$_2$ a$_3$ [$_{VP}$ [e]$_3$ écrit [e]$_4$ *Jean*$_2$] [e]$_1$ [à Marie]$_4$]

The Italian counterpart to (i) and (ii) are also ill-formed, while they are well-formed in Spanish.

(iv) ??Si conosce il giorno in cui ha scritto Gianni a Maria.
(v) ??Mi domando quando ha scritto Gianni a Maria.
(vi) Conocemos el día en que escribió Juan a María.
(vii) Me pregunto cuando escribió Juan a María.

[11] The indefinite object preceded by *a* is interpreted as specific, and a relative modifier must be in the indicative mood. If the *a* is lacking, the indefinite object is intepreted as nonspecific, and the relative modifier must be in the subjunctive. Cf. *Juan busca una muchacha que sepa hablar inglés* (J. seeks a girl that can speak (subj) English.).

[12] Note that *a* in this example is not a preposition, as shown by the possibility of cliticizing the object with the ACC clitic *lo*: *El sujeto **lo** precede/sigue.*

[13] We assume that temporal adverbs may be attached either to the T_V projection or to the T_N projection. This is a necessary assumption to account for the possibility of temporal adverbs preceding the postverbal subject in Spanish. See (26) and (27).

[14] This point can also be illustrated with the post-nominal possessors in Romance. The possessor of alienable nouns like *book* and *table* can be argued to be in the Spec of D (See Zubizarreta & Vergnaud (1992)). When the possessor is a lexical NP, it appears in post-nominal position (non-adjacent to D); e.g. *de María* in *el libro de María* 'the book of M.' This is possible to the extent that D is an N-related category.

[15] Note that given this analysis of the sentences with postverbal subject in Italian and French, the constraint on PP extraction in (38), (42), (43b), (44b) can no longer be explained in terms of the Crossing Constraint. An alternative explanation must be sought.

[16] Adverb preposing in the case of infinitival adjuncts is not possible either (although these can take a lexical subject as noted earlier): **Al temprano salir Juana de casa,...* (When early leave(inf) J. the house,...). This must be due to independent reasons. In fact nothing can intervene between the temporal complementizer and the verb (not even an aspectual adverb). The complementizer and the verb seem to form a syntactic unit in these structures.

[17] Adjunction to infinitival clauses is not possible either. Thus, as Piera notes, the adverb has scope over the entire clause in the following example:

(i) En esta ciudad, salir de casa es malo para salud.
'In this city, to leave the house is bad for one's health'

and the following sentence is ungrammatical:

(ii) *Cansado, salir de casa es malo para la salud.
'Tired, to leave the house is bad for one health.'

[18] (67b) becomes possible if the fronted object bears contrastive stress. Hernanz & Brucart (1987) have shown the in such focused structures the fronted element is in Comp: it has all the properties of wh-movement (it is an unbounded movement, which obeys subjacency).

[19] A sentence like *Nadie no llamó* 'Nobody not called' may only be interpreted as double negation (i.e. as equivalent to *Alguién llamó* 'Someone called').

[20] While the clitic may not be easily dropped in (i), it may be dropped in (ii).

(i) ??*A nadie* examinó/visitó/vió el médico.
(Noone(DO) examined/saw/visited the doctor(SUBJ)

(ii) *A nadie* saludó el médico.
(Noone(DO) greeted the doctor(SUBJ))

A difference between the verbs in (i) and in (ii) is that *examinar/visitar/ver* are obligatorily transitive and *saludar* is optionally transitive. The above contrast might be linked to this distinction.

[21] The sentences in (79) must not be confused with focalization (see note 18): if the preposed negative polarity item bears contrastive stress, the sentences become acceptable.

[22] Rizzi suggests that the root/embedded asymmetry might be attributed to the fact that subject inversion is optional in indirect questions with an embedded subjunctive . So there might be some interference from the subjunctive when judging the indicative case. There are no genuine subjunctives in indirect questions in Spanish.

REFERENCES

Aoun, J, & A. Li: 1993, *Syntax of Scope*, MIT Press.

Belletti, A. & L. Rizzi: 1988, 'Psych Verbs and θ-Theory,' *Natural Language and Linguistic Theory* 6, 291-352.

Belletti, A. & L. Rizzi: 1981, 'The Syntax of 'ne': Some theoretical implications,' *The Linguistic Review* 1, 117-154.

Borderlois, Y.: 1974, *The Grammar of Spanish Causative Complements*, Doctoral Dissertation, MIT.

Chomsky, N. & H. Lasnik: 1991, 'Principles and Parameters Theory,' In J. Jacobs, A. von Stechow, W. Sternfeld, and T. Vennemann, eds., *Syntax: An International Handbook of Contemporary Research*, Walter de Gruyter, Berlin.

Chomsky, N.: 1992, 'A Minimalist Program for Linguistic Theory,' ms., Linguistics Dept., MIT, Cambridge, Mass.

Contreras, H.: 1991, 'On the Position of Subjects,' In S. Rothstein, eds., *Perspectives on Phrase-Structure: Heads and Licensing, Syntax and Semantics* 25, Academic Press, San Diego, 61-79.

Elordieta, G.: 1992, 'Lexical Subjects in Spanish Infinitival Adjuncts,' unpublished ms., Linguistics Dept., USC, Los Angeles.

Franco, J.: 1992, 'On Object Agreement in Spanish,' unpublished ms., Dept. of Spanish & Portuguese, USC, Los Angeles.

Goodall, G.: 1991, 'On the Status of Spec of IP', paper presented at WCCFL X.

Hernanz, M.L. & J.M. Brucart: 1987, *La Sintaxis(1)*, Editorial Crítica, Barcelona.

Kayne, R.: 1972, 'Subject inversion in French interrogatives', In J. Casagrande and Bohdan Saciuk, eds., *Generative Studies in Romance Language*, Newbury House Publishers Inc., Rowley, Mass.

Kayne, R.: 1981, 'ECP Extensions', *Linguistic Inquiry* 12, 93-134.

Kayne, R.: 1986, 'Connexité et l'inversion du sujet', In M. Ronat and D. Couquaux, eds., *La Grammaire Modulaire*, Editions de Minuit, Paris, 127-147.

Koopman, H. & Sportiche, D.: 1991, 'The Position of Subjects', *Lingua*.

Lois, X.: 1989, *Aspects de la Grammaire de l'Espagnol et Théorie de la Grammaire*, thèse de doctorat, Université Paris 8, St-Denis.

Mahajan, A.: 1990, *The A/A-bar Distinction and Movement Theory*, Doctoral Dissertation, MIT, Cambridge, Mass.

Meyer, P.: 1972, 'Some Observations on Constituent Order in Spanish', In J. Casagrande & B. Saciuk, eds., *Generative Studies in Romance Languages*, Newbury House Publishers, Inc., Rowley, Mass.

Piera, C.: 1987, 'Sobre la Estructura de las Cláusulas de Infinitivo', In V. Demonte & M. Fernández Lagunilla, eds., *Sintaxis de las Lenguas Románicas*, El Arquero, Madrid, pp. 148-166.

Pesetsky, D.: 1982, *Paths and Categories*, unpublished PhD dissertation, MIT, Cambridge, Mass.

Pollock, J.Y. 1986, "Sur la syntaxe de 'en' et le paramètre du sujet nul", in M.Ronat and D. Couquaux (eds) *La Grammaire Modulaire*, Editions de Minuit, Paris, pp. 211-246.

Pollock, J. Y.: 1989, "Verb Movement, UG and the Structure of IP", *Linguistic Inquiry* 20, pp. 365-424.

Rigau, G.: 1992, "La Legitimación del Sujeto de las Construcctiones Temporales de Infinitivo", presented at the II Colloquium on Generative Grammar, Universidad del País Vasco, Vitoria-Gasteiz, Spain.

Rizzi, L.: 1987, "Three Issues in Romance Dialectology", presented at the GLOW Workshop on Dialectology, 10th GLOW Colloquium, Venice.

Rizzi, L.: 1990, *Relativized Minimality*, MIT Press.

Rizzi, L.: 1991, "Residual Verb Second and the Wh-Criterion", unpublished ms., Université de Genève, Genève.

Roberts, I.: 1991, *Verbs and Diachronic Syntax*, to be published by Cambridge University Press, Cambridge.

Suñer, M.: 1988, "The Role of Agreement in Clitic-Doubled Constructions", *Natural Language and Linguistic Theory* 6, pp. 391-434.

Zubizarreta, M.L. *On the Relation of the Lexicon to Syntax*, unpublished PhD dissertation, MIT, Cambridge, Mass.

Zubizarreta, M.L. & J.R. Vergnaud: 1992, "The Definite Determiner and the Inalienable Constructions in French and in English", *Linguistic Inquiry* 23.4.

INDEX

Studies in Natural Language and Linguistic Theory

Managing Editors

Publications

1. L. Burzio: *Italian Syntax.* A Government-binding Approach. 1986. ISBN Hb 90-277-2014-2; Pb 90-277-2015-0
2. W.D. Davies: *Choctaw Verb Agreement and Universal Grammar.* 1986. ISBN Hb 90-277-2065-7; Pb 90-277-2142-4
3. K. É. Kiss: *Configurationality in Hungarian.* 1987. ISBN Hb 90-277-1907-1; Pb 90-277-2456-3
4. D. Pulleyblank: *Tone in Lexical Phonology.* 1986. ISBN Hb 90-277-2123-8; Pb 90-277-2124-6
5. L. Hellan and K. K. Christensen: *Topics in Scandinavian Syntax.* 1986. ISBN Hb 90-277-2166-1; Pb 90-277-2167-X
6. K. P. Mohanan: *The Theory of Lexical Phonology.* 1986. ISBN Hb 90-277-2226-9; Pb 90-277-2227-7
7. J. L. Aissen: *Tzotzil Clause Structure.* 1987. ISBN Hb 90-277-2365-6; Pb 90-277-2441-5
8. T. Gunji: *Japanese Phrase Structure Grammar.* A Unification-based Approach. 1987. ISBN 1-55608-020-4
9. W. U. Wurzel: *Inflectional Morphology and Naturalness.* 1989 ISBN Hb 1-55608-025-5; Pb 1-55608-026-3
10. C. Neidle: *The Role of Case in Russian Syntax.* 1988 ISBN 1-55608-042-5
11. C. Lefebvre and P. Muysken: *Mixed Categories.* Nominalizations in Quechua. 1988. ISBN Hb 1-55608-050-6; Pb 1-55608-051-4
12. K. Michelson: *A Comparative Study of Lake-Iroquoian Accent.* 1988 ISBN 1-55608-054-9
13. K. Zagona: *Verb Phrase Syntax.* A Parametric Study of English and Spanish. 1988 ISBN Hb 1-55608-064-6; Pb 1-55608-065-4
14. R. Hendrick: *Anaphora in Celtic and Universal Grammar.* 1988 ISBN 1-55608-066-2
15. O. Jaeggli and K.J. Safir (eds.): *The Null Subject Parameter.* 1989 ISBN Hb 1-55608-086-7; Pb 1-55608-087-5
16. H. Lasnik: *Essays on Anaphora.* 1989 ISBN Hb 1-55608-090-5; Pb 1-55608-091-3
17. S. Steele: *Agreement and Anti-Agreement.* A Syntax of Luiseño. 1990 ISBN 0-7923-0260-5
18. E. Pearce: *Parameters in Old French Syntax.* Infinitival Complements. 1990 ISBN Hb 0-7923-0432-2; Pb 0-7923-0433-0
19. Y.A. Li: *Order and Constituency in Mandarin Chinese.* 1990 ISBN 0-7923-0500-0

Studies in Natural Language and Linguistic Theory

20. H. Lasnik: *Essays on Restrictiveness and Learnability.* 1990
ISBN 0-7923-0628-7; Pb 0-7923-0629-5
21. M.J. Speas: *Phrase Structure in Natural Language.* 1990
ISBN 0-7923-0755-0; Pb 0-7923-0866-2
22. H. Haider and K. Netter (eds.): *Representation and Derivation in the Theory of Grammar.* 1991 ISBN 0-7923-1150-7
23. J. Simpson: *Warlpiri Morpho-Syntax.* A Lexicalist Approach. 1991
ISBN 0-7923-1292-9
24. C. Georgopoulos: *Syntactic Variables.* Resumptive Pronouns and A' Binding in Palauan. 1991 ISBN 0-7923-1293-7
25. K. Leffel and D. Bouchard (eds.): *Views on Phrase Structure.* 1991
ISBN 0-7923-1295-3
26. C. Tellier: *Licensing Theory and French Parasitic Gaps.* 1991
ISBN 0-7923-1311-9; Pb 0-7923-1323-2
27. S.-Y. Kuroda: *Japanese Syntax and Semantics.* Collected Papers. 1992
ISBN 0-7923-1390-9; Pb 0-7923-1391-7
28. I. Roberts: *Verbs and Diachronic Syntax.* A Comparative History of English and French. 1992 ISBN 0-7923-1705-X
29. A. Fassi Fehri: *Issues in the Structure of Arabic Clauses and Words.* 1993
ISBN 0-7923-2082-4
30. M. Bittner: *Case, Scope, and Binding.* 1994 ISBN 0-7923-2649-0
31. H. Haider, S. Olsen and S. Vikner (eds.): *Studies in Comparative Germanic Syntax.* 1995 ISBN 0-7923-3280-6
32. N. Duffield: *Particles and Projections in Irish Syntax.* 1995
ISBN 0-7923-3550-3; Pb 0-7923-3674-7
33. J. Rooryck and L. Zaring (eds.): *Phrase Structure and the Lexicon.* 1996
ISBN 0-7923-3745-X
34. J. Bayer: *Directionality and Logical Form.* On the Scope of Focusing Particles and Wh-in-situ. 1996 ISBN 0-7923-3752-2
35. R. Freidin (ed.): *Current Issues in Comparative Grammar.* 1996
ISBN 0-7923-3778-6; Pb 0-7923-3779-4
36. C.-T.J. Huang and Y.-H.A. Li (eds.): *New Horizons in Chinese Linguistics.* 1996 ISBN 0-7923-3867-7; Pb 0-7923-3868-5
37. A. Watanabe: *Case Absorption and WH-Agreement.* 1996
ISBN 0-7923-4203-8
38. H. Thráinsson, S.D. Epstein and S. Peter (eds.): *Studies in Comparative Germanic Syntax.* Volume II. 1996 ISBN 0-7923-4215-1
39. C.J.W. Zwart: *Morphosyntax of Verb Movement.* A Minimalist Approach to the Syntax of Dutch. 1997 ISBN 0-7923-4263-1; Pb 0-7923-4264-X
40. T. Siloni: *Noun Phrases and Nominalizations.* The Syntax of DPs. 1997
ISBN 0-7923-4608-4
41. B.S. Vance: *Syntactic Change in Medieval French.* 1997 ISBN 0-7923-4669-6
42. G. Müller: *Incomplete Category Fronting.* A Derivational Approach to Remnant Movement in German. 1998 ISBN 0-7923-4837-0

Studies in Natural Language and Linguistic Theory

43. A. Alexiadou, G. Horrocks and M. Stavrou (eds.): *Studies in Greek Syntax.* 1998 ISBN 0-7923-5290-4
44. R. Sybesma: *The Mandarin VP.* 1999 ISBN 0-7923-5462-1
45. K. Johnson and I. Roberts (eds.): *Beyond Principles and Parameters.* Essays in Memory of Osvaldo Jaeggli. 1999 ISBN 0-7923-5501-6

Kluwer Academic Publishers – Dordrecht / Boston / London